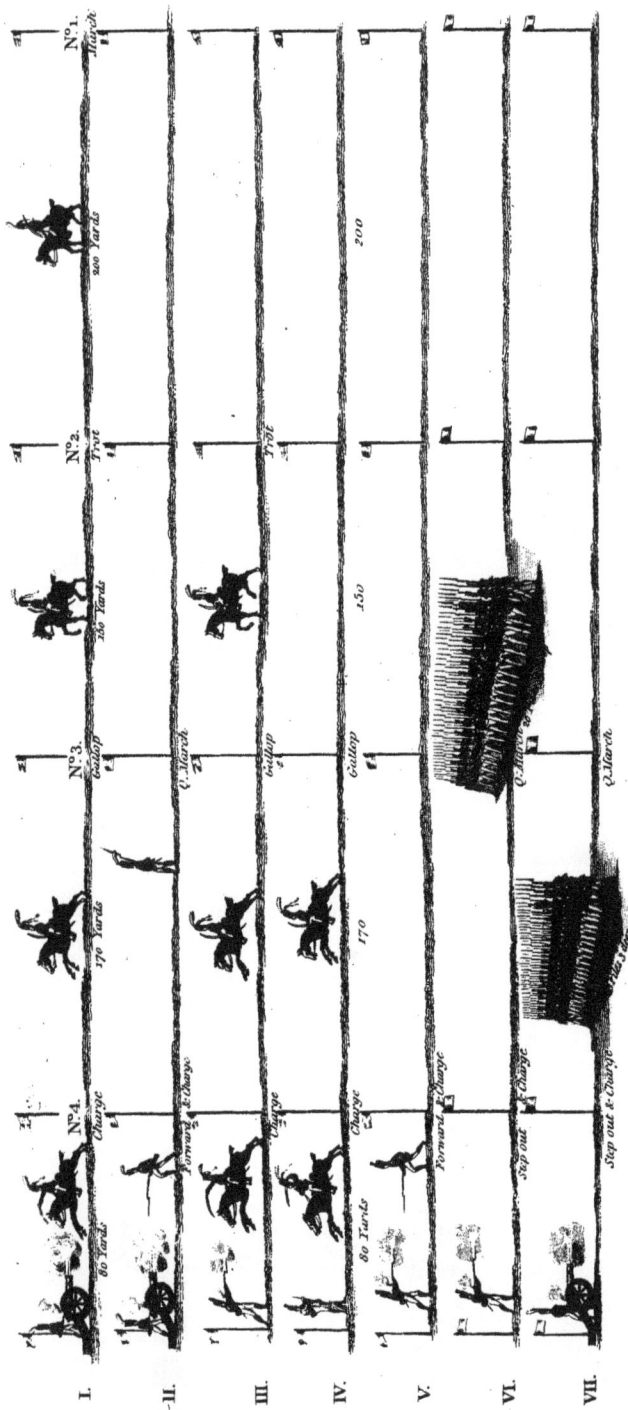

Experiments

Charge of Light Cavalry on Artillery   600 Yards or 720 paces   Nº of Discharges 13   Time 115 Seconds

I.	Infantry attacking Artillery	250	or 300	13	102
II.	Cavalry charging Infantry	400	or 480	3	49
III.	Feint of Cavalry on Infantry	250	or 300		
IV.	Infantry attacking Infantry	250	or "	5	90
V.	Dº dº	250	or "	7	145
VI.	Infantry attacking Artillery	250	or "	14	117

Nº 1. March
Nº 2. Trot
Nº 3. Gallop
Nº 4. Charge

200 Yards
200

150

170 Yards
170

80 Yards
80 Yards

C. March
Gallop
Forward & Charge
Charge
Step out
Step out & Charge & Full Gallop

I.
II.
III.
IV.
V.
VI.
VII.

A SERIES OF
MILITARY EXPERIMENTS

OF ATTACK AND DEFENCE,

Made in Hyde Park, in 1802, under the Sanction of

HIS ROYAL HIGHNESS

THE COMMANDER IN CHIEF.

with

𝕴𝖓𝖋𝖆𝖓𝖙𝖗𝖞, 𝕮𝖆𝖛𝖆𝖑𝖗𝖞, 𝖆𝖓𝖉 𝕬𝖗𝖙𝖎𝖑𝖑𝖊𝖗𝖞;

AND

In the Island of Jersey, in 1805,

BY PERMISSION OF

LIEUTENANT-GENERAL ANDREW GORDON.

WITH NOTES, REMARKS, AND ILLUSTRATIONS.

LIEUT. JOHN RUSSELL, 96th REGT.

Author of Instructions for the Drill, and of Movements of a
Battalion of Infantry, &c. &c.

The Naval & Military Press Ltd

published in association with

ROYAL
ARMOURIES

Published by
The Naval & Military Press Ltd
Unit 10 Ridgewood Industrial Park,
Uckfield, East Sussex,
TN22 5QE England
Tel: +44 (0) 1825 749494
Fax: +44 (0) 1825 765701
www.naval-military-press.com

in association with

ROYAL
ARMOURIES

The Naval & Military
Press

MILITARY HISTORY AT YOUR
FINGERTIPS

... a unique and expanding series of reference works

Working in collaboration with the foremost
regiments and institutions, as well as acknowledged
experts in their field, N&MP have assembled a
formidable array of titles including technologically
advanced CD-ROMs and facsimile reprints of
impossible-to-find rarities.

INTRODUCTION.

THE object of the following Experiments was to ascertain in what time infantry or cavalry, from given distances, and at the regulated military paces, could arrive at infantry or artillery posted to receive them; and also to ascertain how many discharges they might in such time be liable to receive; and it was presumed, that from the result of these Experiments some data might be obtained that might hereafter prove of advantage to the British Service, and it was with a view, no doubt, to this object, that His Royal Highness the Commander in Chief was graciously pleased to order whatever had been required for making the Experiments in

Hyde-

Hyde-Park. Lieutenant-General Gordon, the commander in chief in Jersey, was pleased to order the means for trying similar Experiments in that island : others were intended to have been tried there, had circumstances favoured, particularly with respect to the ranges of grape-shot; but it is hoped, that what has been done, may serve as a ground-work, or as hints at least for others to improve on. On each of the Experiments made, a few Remarks are given; to which are subjoined, some Notes selected from the best military writers. It need only be added, that the whole of the Experiments were made with scrupulous accuracy, with men highly disciplined, in the presence of officers of great professional knowledge; and their results are here faithfully reported.

SUBSCRIBERS NAMES.

A.

Lieutenant-Colonel Adam, 21st Regiment or Scotch Fuzileers.
Captain Austen, 58th Regiment.
Captain Allot, 58th Regiment.
Ensign Alston, 3d Guards.
Ensign J. Armstrong, 66th Regiment.

B.

Major-General the Earl of Banbury, 3d Guards.
Brigade-Major Broad (Captain, Royals).
Lieutenant-Colonel Baird, 83d Regiment.
Captain Blacquaire, 83d Regiment.
Major Blommart, 62d Regiment.
Captain Ball, 62d Regiment.
Captain Beard, 66th Regiment.
Major Boydell, Royal East London Militia, 2 copies.
Lieutenant Bamfurd, Adjutant Light Horse Volunteers.
Major Blackwell, 1st West India Regiment.
Captain Baldwyne, 58th Regiment.
Lieutenant Barnell, Royal West London Militia.
Lieutenant James Barnell, Royal West London Militia.
Surgeon William Box, Royal West London Militia.
Captain Buller, Bank of England Volunteers (Bank Director).
Capt. Bowden, Bank of England Volunteers (Bank Director).

Major-

C.

Major-General Calvert, Adjutant-General to the Forces.
Brigadier-General Carmichael.
Captain Carr, 83d Regiment.
Captain Carroll, Royal East London Militia.
Ensign De Coche, Royal East London Militia.
Lieutenant Crow, 58th Regiment.
Lt.-Col. Le Couteur, Assistant Quarter-Master-Gen. in Jersey
James Calder, Esq.
Adjutant Cudbertson, 96th Regiment.
Ensign Cullymore, 58th Regiment.
Ensign Longfield M'Carty, 8th West India Regiment.
Captain T. H. Cooper, 56th Regiment.

D.

Lt.-Gen. G. Don, Lt. Gov. and Commander in Chief, Jersey:
Colonel Dorrien, Blues.
Lieutenant-Colonel Sir J. Douglas, Royal Marines.
Lieutenant-Colonel Doyle, 87th Regiment.
Captain Donelan, 58th Regiment.
Captain Dominicus, Adjutant 2d Regt. E. I. Volunteers.
William Dorehill, Esq. Chester Place.
Paymaster J. Deans, Royal East London Militia.
Lieutenant Drury, Royal West London Militia.
Ensign Duncan, Royal West London Militia.
Lieutenant Dudgeon, Royals.
Capt. Dorien, Bank of England Volunteers (Bank Director).

E.

George Ellis, Esq. War-Office.
Captain Edwards, Royal West London Militia.
Captain J. English, 66th Regiment.

F.

Lieutenant-Colonel Fitzgerald, 58th Regiment.

James Farrell, Esq. 6th Loyal London Volunteer Infantry.

Captain Franklin, Royal Artillery, Colchester.

Ensign French, 96th Regiment.

Lieutenant Farmer, 58th Regiment.

Ensign Fowle, 58th Regiment.

Captain Forster, Royal East London Militia.

G.

Lieutenant-General Gordon.

Lieutenant-Colonel William Gordon, 83d Regiment.

William Gordon, Esq. Harley Street.

Captain Gomersal, 58th Regiment.

Captain Gage, Royal West London Militia.

Lieutenant Gear, 2d Regiment East India Volunteers.

Captain Gardner, 3d Foot, or Buffs; Aide-de-Camp.

H.

General the Earl of Harrington, Comm. in Chief in Ireland.

Hon. Gen. Harcourt, Governor of the Royal Military College.

Brigadier-General Housten, 58th Regiment.

Joseph Higginson, Esq. Oakfield.

Captain Higginson, Royal West London Militia, 2 copies.

Captain Harris, Royal West London Militia.

Ensign Hellier, Royal West London Militia.

Paymaster Harvey, Royal West London Militia.

Ensign Holder, Royal West London Militia.

Major Hunter, Royals.

Captain and Adjutant Hitchcock, Royal West London Militia.

Lieutenant Hopkins, 2d Life Guards.

Major Henny, 58th Regiment.

Capt. Harman, Bank of England Volunteers (Bank Director).

Capt. Holford, Bank of England Volunteers (Bank Director).

Major-

J.

Major-General Johnson, Guards.
Brigade-Major Johnson, Guards.
Major Charles James, Royal Waggon Corps.
William St. John, Esq.

K.

Field Marshal His Royal Highness the Duke of Kent, 4 copies.
Captain Kerr, 62d Regiment.
Lieutenant and Adjutant Knight, 66th Regiment.
Ensign Kingsly, 58th Regiment.
Ensign Kirk, 3d Garrison Battalion.

L.

Lieutenant-Colonel Leathes, 96th Regiment.
John Lamalet, Esq. Winchester Street.
Lieutenant Lord, Royal Artillery, Jersey ; Aide-de-Camp.
Captain Lovett, Royal West London Militia.
Capt. Langley, Bank of England Volunteers (Bank Director).
————— Library, 21st Regiment, or Scotch Fuzileers.

M.

Captain M'Veal, 66th Regiment.
Captain Malorti de Martemont.
Ensign Maunder, Royal West London Militia.
Lieutenant Morrison, 58th Regiment.
Surgeon Matthews, 3d Foot or Buffs.
Col. Manning, Bank of England Volunteers (Bank Director).
Major Mellish, Bank of England Volunteers (Bank Director).

N,

Colonel N. Newnham, Royal West London Militia.

Captain

O.

Captain Ogilvie, Royals.

Captain Oliver, Bank of England Volunteers (Bank Director).

P.

Major Porter, Royal West London Militia.

Lieutenant T. Pardy, 66th Regiment.

Lieutenant Parsons, Royal West London Militia.

Lieutenant Pearse, Royal West London Militia.

Lieutenant Palmer, Royal West London Militia.

Ensign Parker, 58th Regiment.

Captain Pohlman, 96th Regiment.

Captain Perring, 96th Regiment.

Captain Plenderleath, 60th Regiment.

F. T. Power, Esq.

Q.

Edward Quin, Esq.

R.

Brigade Major Reide, 21st Regiment.

Captain Robinson, Royal West London Militia.

Lieutenant Robinson, Royal West London Militia.

Lieutenant Redmond, Royal West London Militia.

Ensign Russell, Royal West London Militia.

Lieutenant Rogers, 58th Regiment.

Daniel O'Ryan, Esq. Winchester Street.

Captain B. Reynolds.

S.

Captain Stone, 58th Regiment.

Lieutenant Stretch, 87th Regiment.

Captain Simpson, Royal West London Militia.

Ensign Shaftoe, Royal West London Militia.

P. H. Savage, Esq. 1st Life Guards.

Lieu-

T.

Peter M'Taggert, Esq. Loyal London Volunteer Infantry.
Captain Tatham, Royal West London Militia.
Lieutenant Tarratt, R. W. L. Militia (Assistant Surgeon).
Major Thorley, 96th Regiment.
Capt. Thompson, Bank of England Volunteers (Bank Director).

V.

Captain Vaughan, Royal West London Militia.

W.

Brigadier General Wynyard, Deputy Adjutant General,
Major Watts, 3d Roy. Vet. Bn. Barrack Master, Jersey.
Lieutenant-Colonel Wigan, Royal West London Militia.
Francis White, Esq. Winchester Street.
Henry Williams, Esq. Winchester Street.
Captain Whitemore, Adjutant 2d L. L. Vol. Infantry.
Ensign Wolsencroft, Royal West London Militia.

Y.

Major Young, 58th Regiment.

EXPERIMENTS,

&c. &c. &c.

A FIELD-PIECE with artillery gunners, a dragoon of the 11th regiment, and two soldiers of the Coldstream Foot Guards, with two drill serjeants, and the requisite number of camp-colour men, being assembled, in Hyde-Park, at the hour appointed, the ground from the spot where the gun was placed was first accurately measured, and the distances marked as follows, by camp colours, numbered 1, 2, 3, and 4.

(A.)

No. 1 at 600 yards or 720 paces of 30 inches from the gun.
No. 2 at 400 yards or 480　　do.　　　do.
No. 3 at 250 yards or 300　　do.　　　do.
No. 4 at 80 yards or 97　　do.　　　do.

The

The different paces of the horse were then measured in the following manner.

(B.)

The dragoon placed at No. 1, was instructed at the word *March*, to put his horse to as fast a walk as he could to No. 2, and then halt. Trot to No. 3, halt. Gallop to No. 4, halt. Charge to gun, halt. The result was as follows:

(C.)

	Yds.		Sec.
From No. 1 march to No. 2, distance	200	time	95
From No. 2 trot to No. 3, distance	150	do.	28
From No. 3 gallop to No. 4, distance	170	do.	13
From No. 4 charge to gun, distance	80	do.	8
Total yards	600	Seconds	144

(D.)

The several times were accurately measured by a stop-watch, and the dragoon having been particularly cautioned to consider himself, in all his paces, as if acting in squadron, the first experiment was made.

1st EX-

1st EXPERIMENT. (E.)

Charge of Light Cavalry on Artillery.

THE dragoon placed at camp colour No. 1, received the word *March*; at No. 2, *Trot;* at No. 3, *Gallop;* at No. 4, *Charge.*

The dragoon received his several words at the camp-colours, as already stated, and went through his different paces without making any halt, till he arrived at the gun, which he did in 115 seconds; in which time the gun fired thirteen times, the last when the dragoon was within six or seven yards of it.

2d EXPERIMENT. (F.)

Infantry attacking Artillery.

A SOLDIER placed at flag No. 3, received the word *Quick March* to No. 4, distance 170 yards, or 204 paces of 30 inches each ; at No. 4, *Forward and Charge to Gun*, 80 yards, or 87 long paces of 33 inches each. He reached the gun in 102 seconds, in which time it fired thirteen times; at the last discharge, the soldier was almost close to the gun.

3d EX-

3d EXPERIMENT. (G.)

Light Cavalry attacking Infantry.

THE dragoon placed at camp-colour No. 2 (400 yards distance from the soldier), received the word *Trot*, to No. 3; *Gallop*, to No. 4; *Charge*, to soldier. The soldier fired three times; the last, when the dragoon was within about ten yards of him. Time 49 seconds.

4th EX-

4th EXPERIMENT. (H.)

Feint of Cavalry to draw the Infantry's Fire,
Wheel up and Charge.

THE dragoon placed at camp-colour
No. 3, at 250 yards from the soldier, received the word *Gallop;* when arrived near
to No. 4, at 80 yards from the soldier,
perceiving that the infantry waited with
shouldered arms, he increased his speed;
the soldier makes ready; dragoon, as if
intimidated, suddenly wheels off; infantry,
supposing the cavalry to have fled, give
their whole fire; dragoon instantly wheels
up, and from No. 4, charges with rapidity.
He arrived at the soldier before he could
fire. This experiment was repeated with
a different soldier, and the result was the
same; after giving their fire at 80 yards,
they had just time to load again, but not to
fire.

5th EX-

5th EXPERIMENT. (I.)

Attack of Infantry against Infantry in Line.

THE attacking soldier was placed at camp-colour No. 3, at 250 yards from the defending soldier; at the word *Quick March,* the attacking man stepped off; the other soldier instantly began to fire. When the attacking man arrived at No. 4, he received the words (from the drill serjeant who accompanied him all along) *Forward—Charge.* When arrived at defending man, *Halt.*

Quick March from flag No. 3 to No. 4 (170 yards, or 204 paces each of 30 inches), was made in 70 seconds. *Charge* from No. 4, in 20 seconds ; 80 yards or 87 paces

of

of 33 inches.   Total time, 90 seconds;
total yards, 250.

In this time the defending man fired five
times; at the last discharge, the attacking
man was close to him with his charged
bayonet.

━━━━━━━━━

6th EXPERIMENT. (K.)

Jersey, Feb. 10, 1805.

THE two following Experiments were
made in Jersey, by permission of Lieutenant-General Gordon, Commander in
Chief, &c. &c. &c.

In this Experiment made at Saumarez
Miles, near St. Héliers, the camp-colours
were

were placed as in the Experiments made at Hyde-Park.

A soldier of the 1st Battalion of the 83d Regiment, commanded by Lieutenant-Colonel Baird, fully accoutred, was supposed to represent a posted enemy. Sixty men of the same battalion, formed three deep, and with their proper officers, were placed at Flag No. 3, with their centre directed on Flag No. 4, and on the soldier, and a flag placed in a line with him, and beyond him.

At the first word *Quick March*, the party stepped off, and the soldier at the same instant began to fire; at Flag No. 4, they received the words *Forward* and *Charge*; when at about twenty yards from the soldier, *Run—Halt*. The result was as follows:

			Seconds.
From No. 3 to No. 4	170 yards, or	204 paces, in	100
From No. 4 to Soldier	80 yards, or	96 paces, in	45
	Yards 250	Paces 300	Time 145

were

In the 145 seconds, the soldier fired seven times; the last, when the party was within two paces of him.

Lieutenant-Colonel Baird gave the several words of command, with his usual precision. The ground was very uneven, rushy, and at No. 4 so wet, that the officers were splashing mid-leg in water; to such drilled men this made no difference; their dressing was exactly preserved the whole of the way; no anxiety was shewn by officers or men, but to have the experiment made as satisfactorily as possible. Captain Carr of the same regiment held the stop-watch.

7th EX-

7th EXPERIMENT, (L.)

Jersey, April 10, 1805.

THIS Experiment was made in the presence of Lieutenant-General Gordon, on a fine hard strand, at low water, at the Miles, near St. Héliers. The colours were placed at the several distances, as in the last Experiment. A six-pounder, with the requisite number of artillery men, &c. under the immediate direction of Lieutenant Barlow of the Royal Artillery, was posted to fire at 80 yards from Flag, No. 4.

The Light Company of the 83d Regiment, with its officers, and formed three deep, was placed with its right flank at Flag No. 3, directed on Flag No. 2, and the gun, keeping it however a little open, for fear of any accident.

Lieu-

Lieutenant-Colonel Baird gave the several words; At the word *Quick March,* the men stepped off; and at the same instant the gun began to fire. The result was as follows:

<table>
<tr><td></td><td></td><td>*Seconds.*</td></tr>
<tr><td>From No. 3 to No. 4 the quick march was made in</td><td>100</td></tr>
<tr><td>From No. 4 to Gun, Step Out, Charge, Run, Halt,</td><td>17</td></tr>
<tr><td></td><td>Total 117</td></tr>
</table>

In this time the gun fired fourteen times; at the last discharge, the men were close to it. The stop-watch was held by Ensign and Adjutant Dorhill of the 83d Regiment.

In the whole of the march of the Light Company, not a man, even at their nearest approach to the gun, had deviated an atom from the line; at the word *Halt,* they were in perfect dress; no less indeed could be expected from a flank company of a battalion so highly drilled, and so ably commanded.

To say any thing of the merits of the artillery

artillery would be unnecessary; to fire fourteen times in one minute and. fifty-seven seconds, is of itself sufficient praise.

———————

Grouville Barracks, Jersey,
30th March, 1805.

EXPERIMENT

To try how quick a Man could fire Ball Cartridge; and whether there was any difference in this respect between a Man firing with his Pack on, and in marching order, or one accoutred as for a common Field-Day.

FOR this purpose two men of the Light Company of the 2d Battalion of the 18th Foot, were granted by Lieutenant-Colonel O'Doherty; they were fully accoutred, and with fixed bayonets; one had

his

his pack on, and in complete marching order. They were placed at a little distance from each other on the beach, and a target set up at 100 paces from them to fire at; they were cautioned to come to a proper level each time, but not to take particular pains in taking aim at the bull's eye, but to load and fire as fast as they could, taking care only to ram their charges well down.

The first fire was given by word of command from Captain Bird, who commanded the Light Company, after which no word was given.

The first eighteen rounds were fired in five minutes. By some accident it happened, that the men had no more than eighteen rounds each : more was therefore brought, and in the mean time the firelocks had time to cool. Ammunition was then supplied by two serjeants, who put it into the pouches, or gave it by hand as wanted, which enabled them to fire much

quicker

quicker than before, near one fourth : but as the object was to ascertain, precisely, how quick they could fire, by taking the cartridges themselves from their pouches, and how long they could continue it ; so no calculation could be safely made, but on the first eighteen rounds, on account of the interruption that happened in getting more ammunition. At the thirty-seventh discharge (including the first eighteen), the barrels of the pieces became too hot to be touched by the hand; one man having for some time held his piece by the sling while loading ; and the other, whose sling had got loose, held by the swell of the stock; it was therefore thought advisable to desist.

The man in marching order fired thirty-five times, while the man without his pack fired thirty-seven times.

A great number of shots struck the target; all were very near it.

Lieutenant Phillips held the stop-watch.

The two men who fired, were taken at random from 120 of as fine young men as ever stood on a parade; this company in discipline, particularly in the essential matters of firing and manœuvre, were excelled by none.

—————————

St. Hélier, Jersey,
May 29, 1805.

EXPERIMENT

To try in how short a time a Man could fire Thirty-six Rounds of Ball Cartridge.

LIEUT. COL. FITZGERALD commanding the 2d Battalion of the 58th Regiment, in compliance with my request, ordered the Adjutant to furnish me with whatever I might require for the purpose of trying this experiment.

A soldier

A soldier was selected with thirty-six rounds of ball cartridge in his pouch; he was the flugel man of the regiment, and had been many years a soldier.

A target to fire at was set up at 121 paces from the soldier.

The first discharge was made by word of command; after which, the man went on to load and fire as fast as he could.

The first three rounds were fired in one minute; the remainder of eighteen rounds in five minutes and a half; the whole of the eighteen in six minutes and a half. And deducting the time lost in turning the cartouch box, which the soldier could not do without being assisted, the thirty-six rounds were fired in thirteen minutes. After twenty-five discharges, the firelock became too hot to be held in loading, except by the sling. Fifteen balls were put into the target, and the rest were very near it. After every discharge, the fire-

c   lock

lock was properly loaded, and the cartridge well rammed down, with two smart
strokes of the ramrod every time.

Ensign and Adjutant Shea held the stop
watch.

* Whether men in the ranks, especially in the center or
rear ones, could have fired so quick, or so well as men acting independently, as the soldiers in the Experiments here
made, may be doubted. No favourable opportunity, however,
occurred of trying it, as could have been wished. A mode
of ascertaining this by experiment, that would have put it beyond doubt, as well also as to prove the value of the fire of the
rear ranks, has been submitted since, in a small pamphlet, entitled, REMARKS ON THE INUTILITY OF THE THIRD RANK
OF FIRELOCKS.

RE-

RESULTS of the EXPERIMENTS.

ARTILLERY.
HYDE PARK.

No. 1 (E.) fired 13 times in 115 seconds

No. 2 (F.) fired 13 times in 102 seconds

JERSEY.

No. 7 (L.) fired 14 times in 117 seconds

Total 40   334

INFANTRY.
HYDE PARK.

No. 3 (G.) fired 3 times in 49 seconds

No. 5 (I.) fired 5 times in 90 seconds

JERSEY.

No. 6 (K.) fired 7 times in 145 seconds

Total 15   284

IN-

INFANTRY.

(BALL CARTRIDGE.)

JERSEY.

No. 1   fired   18 times   in   5 minutes
No. 2   fired   36 times   in   13 minutes

Total 54   18

REMARKS.

REMARKS.

REMARK (A).

THE Flags No. 1, 2, 3, and 4, were set
up to mark the several distances from
which the infantry and cavalry, in the
following Experiments, were to set out;
and also to mark their several paces, according to principles laid down by Monsieur Guibert, in his admirable work,
L'Essai General de Tactique, where he says,
" That the speed of cavalry" proceeding
to the charge " should be gradual and
progressive:" which he defines " a progressive motion, and which continues to

c 4   increase

increase its celerity till arrived at the enemy." (1)

The same principle is laid down in the Regulations for the British cavalry :--- " That whatever distance the squadron has to go over, it may move at a brisk trot till within 250 yards of the enemy, and then gallop; the word *Charge* (2) is given when within 80 yards (3), and the gallop increased as much as the body can bear in good order;" that is to say, in preserving its dress. And though it might appear that the above extracts relate to cavalry only, yet so analogous are the principles on which infantry and cavalry act, that what is laid down for the one, will be found equally to apply to the other; the charge of both is made on the principle of gradual and increased celerity of movement (4).

Great alterations have taken place in the arms and tactic of the armies of Europe, even since the introduction of gunpowder.

powder. The order of depth was long maintained by generals of great and distinguished abilities (5). Now the thin order is adopted as the best (6). In the first case, troops had the advantage of forming with small fronts, by which it was supposed they could be more easily managed (7). In the other case, they are spread out to great extent, each opposing army endeavouring to out-wing its adversary. The force of the shock of cavalry was placed in its specific weight: large men and heavy horses (8) were therefore selected for the purposes of war; and the heavier they were, the greater it was supposed would be the force of the charge, which was made at no greater pace than a slow trot, both men and horses being covered with armour, and formed six or eight deep. This ponderous body of iron in motion was to bear down all before it by its weight; or, as a tower, resist, immovable, the most violent shocks of the enemy. With a much lighter cavalry formed in thin order (9), the same quantity
tity

tity of weight and effect in the charge is
now supposed to be produced by increased velocity of motion. Thus, contrary to the old tactic, the speed of cavalry
(10) is now esteemed the first and most
useful of its properties; and by the shock
is meant, at this day, the violence with
which a body of horse, put to its full career, can, in dress (11), dash itself against
an opposing body, and overturn it.

REMARK (B).

THOUGH it might not be supposed
possible so to regulate the paces of cavalry (1), that a troop horse, like a foot
soldier, could be brought to step a certain
number of inches each pace, and no more,
and a given number of such paces in a
minute,

minute, yet it cannot be denied, that it is material to know exactly at what rate the squadron can move in any of its paces (2); some one horse in the squadron, and that the slowest one, would necessarily be so placed as to regulate the movement of the whole (3), otherwise different kinds of motion would exist in the squadron at the same time, its uniformity of movement would be destroyed, its dress could not be preserved, and the force of the charge would be consequently lessened, as the enemy could not be attained on all his points at the same time. If this be correct, it is evident that the paces of the leading horse should be proved; and this is what has been attempted here, as a means of obtaining a data on which to found calculations that might be depended on.

Officers and non-commissioned officers of the cavalry should unquestionably be well acquainted with the different paces of their horses, so that they may be
calculated

calculated on at all times, and in all situations, with certainty (4); thus, if the commander of a body of cavalry in a meditated charge on the enemy has perfected himself in his *coup d'œil* (5), he can readily form a correct judgment of his distance from them, in every circumstance, and in all situations of ground; and if he is then certain of the rate at which the squadron can move in all its paces, he may judge, to a great degree of certainty, what time it will take him to attain them, and how many discharges he would in that time be liable to receive from artillery or infantry, whichever arm (6) was opposed to him. This is supposing (what no doubt should be the case), that the cavalry officer should understand, in some degree, the tactic of the infantry; the infantry that of the cavalry; and both of the artillery, at least as to its effects (7); and it would be essential also for the artillery officer to be acquainted with the tactic of the infantry and cavalry. Each arm with us (except in time of war) is
probably

probably kept too separate, and consequently ignorant of each others powers.
The sister arms (cavalry and infantry)
should especially, wherever there was an
opportunity, be frequently exercised together; that having a perfect knowledge
of the benefit and security that each could
afford to the other, of the advantages in
short of mutual assistance and support,
prejudices and jealousies (8), if such exist,
might vanish, and victory attend them in
the day of battle.

RE-

REMARK (C).

THE mounted officers of the infantry should attend very particularly to the paces of their horses, to enable them to take up distances accurately, in all situations, as well as to conduct the column at the regulated step; by calculating the paces of the horse, and applying it to that purpose.

The method of leading the column, whether of infantry or cavalry, or composed of both, whichever arm might lead, has, it is true, been accurately laid down in our tactic, and if strictly attended to must be nearly infallible (1); circumstances might however arise, where every aid would be required to ensure exactness of movement, especially ": when the arrival of the column at a given point is to be perfectly punctual;" and when it is

essential

essential, perhaps, that several divisions of an army should, by different roads, and from several places, arrive, at given points, at a fixed moment of time, where any arriving too soon, or too late, might produce consequences equally fatal. In such critical cases, where every precaution must be used, and where so much depends on preserving the cadenced time, the leading officer depending on the even paces of his horse, might, in the dark and tempestuous night, and in every variety of ground or road, conduct the column with a precision, in such circumstances, impracticable perhaps by any other method.

The necessity of having troops at all times, and on all occasions, habituated to move at the regulated step, can never be too much insisted on ; on it, the exactness of all military movements must depend, and on this precision and excellence of the march must a general rest his hopes of success in war. It is being more perfect in this, that gives a small army the superiority over one more numerous, though composed of men equally courageous and well appointed; it is the essence of discipline, or rather it is discipline itself: the more the subject is considered, the more important it will be found.

It is not the march of a dozen or two paces of a battalion in line, as is practised at reviews, on a verdant turf, level as a bowling green, that is here meant; it 'is the march in line continued for several hundred yards, and in every sort of ground, without any deviations (2); it is the habitual precision of division marching (3), the accurate cadenced step, at which at all times, and on all occasions, the column of route should proceed, so that in a moment the troops might be prepared for battle, without hurry, confusion, or disorder.

The great importance of the accuracy of the route march is too well known to be insisted on here: with a view, no doubt,

to

to this essential object, it is, that the Russians, as we are informed by Tielk, " generally measure the length of each day's journey when the troops are on their march;" (4) an admirable plan, which might however be improved on, and, if adopted by the British army, be highly conducive to the benefit of the service.

REMARK (D.)

AT every flag a proper person was stationed to give the words of command to the several attacking men, and to mark exactly the moment of their arrival at, and their departure from, each flag.

Captain Thorpe, Assist. Dep. Adj. Gen. held the stop-watch.

RE-

REMARK (E).

(1st Experiment.)

Charge of Light Cavalry on Artillery.

THE First Experiment being meant as an endeavour to ascertain in what time a squadron of light cavalry could, from given distances, and at named paces, arrive at a field piece, which it meant to attack; and also to ascertain at the same time, how many discharges the gun could make while the cavalry traversed a certain portion of ground; the flags were placed accordingly: No. 1 at 600 yards from the gun, that being the extremest horizontal range at which it was supposed a light six-pounder could do much execution (1). From this flag, the dragoon, who was supposed to be the leading flank man of the squadron, was to begin his movement. At a signal given (by firing a musket), and

at

at the same instant the cannon fired, and the dragoon began his movement, he went over the 600 yards, at the different paces in 115 seconds, that is in 29 seconds less time, than when his paces were measured: this difference was accounted for by his eagerness to press his horse to the utmost in all his paces, as the gun was firing the whole of the time; and it was perceivable also, that the artillery men exerted themselves with increased anxiety, in proportion as the dragoon approached them, which probably would not have been the case had the one galloped, or the others fired against time. Probably no line of cavalry, or even a squadron, could have gone over the same extent of ground (preserving their order) in so short a time as the horse that was selected for these experiments; who, added to the advantage of independent (2) movement, was a fast goer in all his paces.

It is also a principle laid down in the cavalry tactic, " that horses must not be

brought

brought up blown (3) to the charge."—
For these reasons it might be concluded,
that a body of cavalry moving over 600
yards level ground might receive even
more than thirteen discharges from a field
piece equally well served as the one here
mentioned. At the same time some de-
ductions should be made for new laying
the gun, as its horizontal level in the experiment here made was never altered,
which probably would not have been the
case in action.

The necessity of cavalry and infantry
having some knowledge of artillery, at
least of its effects, has been already mentioned, that they might be aware of how
many shots they might be liable to receive
in traversing a certain portion of ground,
and whether the ground was of such nature as to admit the balls to *ricochet* (4):
and the artillery officer should well understand what degree of rapidity troops
could move at, in every sort of ground,
that he might so regulate his fire as to
deter

deter them from either advancing or retiring (5), and to know almost to a certainty in what time they could arrive at
him, from given distances, in line, or in
the irregular attack *en débandade*. Approached in this latter manner, he would
know that his round shot could do no
great injury to them, and when arrived
within the range of his grape, he could
only depend on this to defeat them, and
then would appear the advantage he must
have in understanding the rate of the enemy's movements; for he would then know,
whether he might venture to fire at such
distance; that though he could do execution, yet that he might have time to load,
so as to make his final discharge when
they were approached so near that he
might be certain of its defeating them,
for that discharge in artillery is generally
destructive and decisive ; or, in short, if
he saw no prospect of giving that fire with
certainty of success, he might abandon
the gun, in time to save his men, for after
the final discharge he has no further means
of defence.

RE-

REMARK (F).

(2d Experiment.)

Infantry attacking Artillery.

THE ancients had the great advantage of forming in line very near the enemy, and it was very essential that they should approach on correct dress as they were opposed man to man, hand to hand, and foot to foot. But since the invention of gunpowder, the destructive fire of cannon, and the length of the ranges, obliges to distant deployments. It is therefore laid down in our tactic for the infantry, " that when a considerable corps of troops (infantry) is to act offensively, it must form in line, at latest within 1200 or 1500 paces of a posted enemy, unless the ground particularly favours, and covers from the fire of artillery, the enfilade of which is what chiefly prevents bodies in

column

column from approaching nearer (1); and that space, under the unceasing protection of their own artillery, troops in line will march over in eighteen minutes" (2).

As the time for performing these Experiments was necessarily limited to a few hours, and the day on which they were tried was uncommonly hot, and as attention was to be paid that the people employed might not be fatigued, lest any Experiment should on that account be incomplete, it was therefore supposed that the ground had so far favoured the infantry that was to attack the artillery, that they had arrived at within three hundred paces of the gun, that is full within its point-blank. The object then of this Experiment being to prove in what time infantry could move over a given portion of ground at the regulated step, and also how many discharges it might be liable in that time to receive from a six-pounder, the soldier, fully accoutred, who was supposed to represent the center man of a division, and

D 4   who

who was attended in all his paces by a drill serjeant of the same regiment), was placed at Flag No. 3, 250 yards or 300 paces from the gun ; the instant the word *Quick March* was given the man stepped off, and the gun began to fire, and so continued till the word *Halt* was given: at the word *Forward* and *Charge*, which he received at Flag No. 4, he moved with rapidity, but without running ; he went over the whole of the ground in 102 seconds, in which time the gun fired thirteen times.

What has been already observed, relative to the movements of cavalry, may apply here ; it is not probable that a line of infantry, preserving its dress, could go over the same portion of ground in the time as the soldier here mentioned, who was young, active, and well drilled, and had the advantage of independent movement : a body of infantry therefore moving over 300 paces of ground would, it may be supposed, receive more than thirteen discharges

discharges from a six-pounder in that time. On the other hand, was a charge made *à la débandade*, from the same distance : as each soldier would run with his utmost speed, it is probable that they would not receive so many as thirteen discharges while they traversed the 250 yards of ground.

REMARK (G).

(3d Experiment.)

Light Cavalry attacking Infantry.

IN this Experiment the dragoon (1) was placed at Flag No. 2, distant 400 yards from the defending man ; at a signal given, the dragoon moved forward, and the soldier fired at the same instant ; the

dragoon

dragoon went over the ground at the different paces in forty-nine seconds, in which time he received three discharges; at the last of which, he was so close to the soldier as to prevent the possibility of his again loading his musket.

From this trial it would appear, that in a charge of light cavalry on infantry, over 400 yards of dry and level ground, and moving at the rate as here stated, the commander might safely calculate on not receiving more than three discharges from each individual of the front rank, and perhaps not so many from the others. The most destructive fires, and the most to be dreaded, are those given by a well disciplined infantry in steady order, as they can apply their fire to its proper purposes, that is to say, to repel an attack: the commander of the infantry therefore being sensible of this, and understanding also the tactic of the cavalry, and having besides a correct *coup d'œil*, so as to judge, at a glance, not only of the distance they were

were at from him, but the nature of the ground that lay between them (which must determine the degree of speed they could approach him at), also whether the horses were fresh or fatigued, he would form his his calculations with the quickness of lightning; and decide, whether he should attempt to give three fires, or only two. He would be aware, that in the first case, his fire given at 400 yards would prove but little executive, and might for that reason dishearten his men; and although he might give two fires more before they could attain him, yet would he rather depend for success on giving but two fires in the whole; the first, when the cavalry were within 200 or 250 yards; and the second, when within twenty-five or thirty of him; or perhaps at fifteen yards: these fires would be better given in vollies of the whole battalion, rather than from its fractional parts. The first of these fires might probably be best given by the four center companies (suppose) direct to the

the front, and the other companies of the
wings firing more or less oblique inwards,
so that the whole weight of this destructive fire might be thrown at once on the
enemy's center. A dreadful chasm would
appear; a void so palpable, and a slaughter so appalling, might deter the stoutest
hearts from proceeding (2); the survivors
would seek for safety in immediate flight:
but if instead of making his fire convergent, the commander of the battalion rather chose to give it along the enemy's
front, he might calculate that it would
knock down or disable a considerable number of the front rank men and horses, be
productive of great and immediate confusion, and consequent delay, as well by the
rear rank stumbling over, or swerving
from, the dead or wounded bodies of the
men and horses that fell in the front rank,
as by the endeavours of the men in the
rear rank to fill up the intervals so made,
that the front rank should be complete,
and in dress, so as to enable it to continue
the charge in its full velocity, and effect,

for

for it is in this rank alone that the force
of the shock is supposed to consist, the
following ranks, how numerous so ever
they might be, adding no force, that is to
say, no weight, to the first. (3)

But whether the first fire was given
oblique, or otherwise, the second general
discharge (4) would, at all events, be made
direct to the front, when they were within
twenty-five paces, or less, of them, as in
this case they would be too near to admit
of a fire being made so oblique (5) as to
be of more use than the direct fire; where
every soldier would be instructed to single
out his man to fire at, and on this the commander would depend for defeating the
enemy; but should this fail, his men would
be prepared to resist and resolutely to receive them on their bayonets, that is to
say, the bayonets of the first rank (for, according to the method used by the different powers of Europe, in arming and instructing the infantry) the bayonets of the
second

second rank are, in this case, of little use,
and those of the third of none at all. (6)

Nothing perhaps would tend more to
give the soldier a confidence in his own
powers, and a relish for his profession,
than making him acquainted with the use
and meaning of every manœuvre (7) in
which he is exercised, and also in what
consisted the fort, or the feeble, of whatever description of troops he might be opposed to; thus, in the instance of a charge
of cavalry, as here supposed, he would
know that if the battalion consisted of 600
men, formed three deep, and in line, that
such would have a front of 200 files, but
that the front of a body of cavalry, covering the same extent of ground, would consist of no more than 128 men, and that
each of these would be exposed to the fire
of nearly five soldiers, that before they
came to the end of their career, it is probable (if the fire of the infantry was of
any value) that they would be diminished
one third, in this case 170 men would be
opposed

opposed to 600, for, notwithstanding the destruction made among the cavalry by the fire of the infantry, yet the latter would not have lost a man; for it is not to be imagined that a well instructed cavalry, as is here supposed, would make use of the irregular fire of carbines or pistols, to shake the ranks of the infantry, for this would be to oppose their feeble to the fort of the others, who could give their fire in a firm position and on a proper level; they would therefore rather apply their fort, the spur and the sabre: in the uniform celerity of the movement, the force of the shock is said to exist, (8) an attempt to fire would therefore weaken and retard it, and destroy its effect; all this the soldier should know, and that should the cavalry even break into the ranks, that miserable indeed must that infantry be, in point of courage, discipline, or bodily strength, two or three of whom, with their fixed bayonets, would not be an over-match for any horseman in Europe, (armed as he now is) however adroit he might be in all the parade of

what

what is called the sword exercise. (9)   But
an alteration in the arming of the battalion, as is here supposed, would still add
to their confidence, and increase their security; the inutility of the third rank of
firelocks, (except perhaps when the front
rank kneels) has been long acknowledged,
and has been endeavoured to be demon-
strated in another place, (10) where it has
been presumed to recommend the substi-
tuting of pikes in their stead; besides the
other advantages arising from this arrange-
ment, the pikemen by supplying the two
ranks in their front with ammunition (11)
would thus increase their fire, and make
it equal in quantity to that of three ranks,
and superior to it in effect; the battalion
then thus armed, the front rank kneeling,
with their bayonets sloped, (which is supposed, though perhaps erroneously, to be
the best position for that rank in opposing
cavalry) (12) the bayonets of the second
rank projected forward to push at the nostrils of the horses, (13) and the spears of
the third rank, reaching beyond these, ele-
vated

vated above them, and aimed at the faces of the horsemen, (14) would present such a barrier, bristling with shining and deadly points of steel, that no cavalry in Europe could attempt to break it down without risk of utter destruction. Let it, however, be for a moment supposed, that the charge was made with such velocity, that the impetuosity of the shock was so great that every thing gave way before it, (15) yet as their ranks must be thinned in the attempt, such as penetrated would be carried away, by the velocity of the motion, beyond the infantry; for the horse torn by the spur, maddening with pain, confused by the whistling of the bullets, and the clamour around him, and feeling himself suddenly freed from the agonizing pressure of the squadron, would hurry his rider far to the rear before his career could be checked. In this case the infantry might recover their temporary confusion, and then would the superiority of the pikes be seen, for the pikemen being then in the rear, would face about, and with protended spears oppose the cavalry,

E　　　　　　who,

who, however exhausted they might be, would be obliged again to prepare for a new attack, when it might be impracticable perhaps to bring the horses to a fresh charge; and thus, instead of having gained a victory, they would themselves be inevitably destroyed; for the firelock men, seeing the steadiness of the pikemen, and the protection they now afforded them, would begin to renew their fire, or to use their bayonets. (16) But if in the first instance the infantry had been dispersed, the horsemen must scatter also and mingle among them, and in this *melée*, as each horseman would have three, or perhaps, four, soldiers, armed with bayonets and spears, to encounter, any horseman that escaped being killed or made prisoner, must have owed his safety to the fleetness of his horse.

Still, however, there is something wanting to give the soldier perfect confidence in himself, some covering for his head in the day of battle; all animals seem instinctively

stinctively to endeavour to guard their
heads from danger; the soldier, conscious
of his defenceless state, shrinks, appalled,
from the arm of the horseman, he sees nothing in idea but the sabre waving over
him, ready to descend, like lightning, on
his unprotected head and cleave him to
the earth; 'tis apprehension of this sort
that fills the minds of the infantry with
such terrors whenever they suppose that
the cavalry can break in on them; but
were the soldier's head defended by a helmet or leather cap, and his shoulders by a
chain epaulette, or *gard d'epaule*, (17) he
would not, in this case, fear singly to oppose a horseman; for instead of the timid
wretch, endeavouring only to save his head
from the edge of the sword, he would
boldly lay himself open, and in his turn
attack the horseman and throw him on his
guard, and consequently defeat him. It
may indeed be affirmed that a battalion,
(armed as is here supposed) so far from
dreading the shock of cavalry, would fear-
lessly oppose themselves to it, and would

E 2   in

in their turn (wherever it was possible from the nature of the ground, and that they could not be out-flanked) attack the cavalry with every prospect of defeating them, for it is laid down in our tactic, that " cavalry has but one method of engaging, which is by charging; if it stands still to receive the shock, its fate is inevitable;" (18) and though the above extract relates to cavalry as attacked by cavalry, yet would it be found equally to hold good were they to be attacked by infantry well disciplined and armed as is here supposed.

REMARK (H.)

Feint of Cavalry to draw the Infantry's Fire,
Wheel up and Charge.

THIS experiment was intended to show
how cavalry might feign an attack, to draw
the fire of the infantry, and, having succeeded, to come on them before they
could again have time to load. It was an
endeavour also to show how dangerous it
would be for infantry to part with their
full fire, at such distance (on the supposi-
tion of the cavalry having fled) as to preclude the possibility of their being again
able to load and fire, before the cavalry
could have time to return to the charge.

It may likewise serve to evince the necessity there must be for infantry and cavalry to understand each others tactic, not

E 3   meaning

meaning the interior and minute detail, but the manœuvre, and how they could affect each other in the day of battle, so that each might be assured of its own powers, as well as those of the attacking arm.

Thus, should the cavalry officer perceive that the infantry commenced their fire when he was at 4 or 500 yards distance, he might naturally be led to suppose that they were unsteady, or shaken in point of courage, or that their commander did not well understand the range of his shot, but threw it away at such distance, that its effect could be but little, and that therefore his attack would be followed by success; but, as he proceeded, if he observed that they kept a fire in reserve, and appeared resolute, to wait his still nearer approach, to pour it on him with more assured destruction, he would, in that case, instantly perceive that his only chance of success must lie in the want of skill in the commander; he would therefore feign to fly, to entice them to give the whole of their fire,

at

at such distance, that he could, if his finesse succeeded, have time to wheel up and charge them, before they could again have time to load and fire, but, on the contrary, if he found that he had failed in drawing their fire as he had hoped, he would then continue his retreat in good earnest, and at his utmost speed.

On the other hand, the infantry officer, finding that his fire given at 4 or 500 yards had no effect, might then reserve his fire till he saw the cavalry so near, that he might hope to give it with better success; but seeing them, when arrived at about 70 or 80 yards of him, suddenly wheel off, as if intimidated, should he, in that case, be induced to give them his whole fire, (1) he might be exposed to certain destruction; his men, astonished and confounded at seeing those return to the charge whom they had supposed to have fled, would be panic-struck and incapable of resistance. The officer who could make such mistake, might, by an endeavour to rectify it, fall

into

into an error still more fatal, by supposing on seeing the cavalry begin their wheel up, that he had still time sufficient to make another discharge, but, in his endeavour to effect it, the cavalry would fall on him in the defenceless condition of loading. It is evident, from what has been here said, that in the disaster supposed to have fallen on the infantry, the fault must have been in their commander, for not having a correct *coup d'œil* to judge of distances, not understanding the ranges of his shot, so as to know the distances at which it could be given with its greatest effect; not knowing exactly how long it would take his men to load, so as to enable them to fire without being hurried or confused, being so ignorant of the tactic of the cavalry, and its movements, as not to be aware of the time in which they could over-run such portion of ground as lay between them, and the time it would take them in the wheel in their return to the charge, and, finally, for having given his concluding fire, before he was certain that the

the cavalry had retired to such distance, that if their retreat was no more than a feint, he had it in his power to load and fire again before they could wheel up and arrive at him.

The necessity of cavalry, artillery, and infantry, being frequently exercised together, must be obvious; in regard to the infantry it is particularly useful. Soldiers should be animated with the sight of cavalry and artillery, and used to the charge of the one, and the fire of the other: " by accustoming the soldier," says Captain Tielk, " to the noise of artillery, much of its effect will be lost. Demonstrate to him that very quick firing is not so destructive as he might conceive, and he will move on with intrepidity." And by familiarizing him to repeated and rapid charges of cavalry he would cease (2) any longer to be intimidated by their formidable appearance, their rapidity of movement, the clangor of the trumpets, the gleaming of the swords, or the thunder of
the

the horse's hoofs. The soldier should also be instructed that notwithstanding the apparent fury of the charge, yet the cavalry might arrive at him in such disorder as to weaken much of its effects; the horse may be timid and his rider courageous, and, on the contrary, the rider may be timid and the horse courageous; (3) that horses may swerve at the very moment when every thing decisive may be expected; that the horse will avoid pushing down the soldier, if possible, or trampling on him when down, but especially the soldier should be made sensible that his safety must lie in his courage, the steadiness of his order, the compactness and closeness of his files, and the imminent danger to which unsteadiness or timidity would expose him. All this, and much more, should be explained to the soldier, to inspire him with confidence in his own powers, and bring him to stand firm and unawed, to resist the attack of cavalry; infantry, brought to this perfection, would be very cautiously approached, or even menaced by that arm, now so

much

much the object of its dread; for it is laid down in the Cavalry Regulations, " that cavalry should never attack infantry, unless this last is either shook in point of courage or disposition." Of all this the cavalry officer would, no doubt, be well aware; yet if there was a necessity, at all events, and at all hazards, to penetrate a line of infantry firm and compact, it might be effected by blind-folding the eyes of the horses of the front rank, putting them to their utmost speed, and urging them with the spur, throw them on the infantry and break down their ranks. (4)

REMARK (I).

(5th Experiment.)

Attack of Infantry on Infantry.

THIS was an endeavour to prove in
what time infantry could move over a
given space, at the regulated step, with
their fire reserved, to the attack of infantry, posted to receive them, and how many
discharges they might, in that time, be
liable to receive from the posted enemy.
250 yards was determined on as the distance for the attacking soldier to move
over in quick time, in his advance on the
posted man, for, according to M. Guibert,
" although the horizontal range of a musket may be computed at 180 toises, or 360
yards, yet where the fire can have any
great effect it is seldom more than at 80
toises, or 160 yards, that is to say, the fire
of infantry as ranged in battle, and in the
tumult of action. Beyond this distance
the

the shot is uncertain." (1) If what is said above be just, the commander of the attacking battalion, would march steadily on, without apprehension of much loss, till he arrived within 300 paces of the enemy, which distance, if his *coup d'œil* was correct, he might readily calculate, and knowing, by experiments fairly made, what number of discharges the enemy could make in that time, proceed accordingly, being only careful not to advance so rapidly, as to throw his men out of dress, or precipitate them breathless on the bayonets of the enemy, as it is in such hurried and imperfect order that the enemy would wish to receive their attack, knowing in this case how little effective it would prove. The commander of the attacking battalion thus being sensible of the extent of the injury he might receive in his approach, and knowing also according to Tielk, " that the fire of cannon or musketry is rather calculated for defence, while the force of an attack consists in the use of the bayonet, (3) and the

celerity

celerity of; the charge," and his object being attack, and not, defence, he consequently would. not be tempted to make use of his fire in advance, or what the Prussians call the attack fire, (4) that is, he would not retard his movement by making a halt to fire, and thus give the enemy an advantage over him; for the enemy being already posted could consequently give more fires, and those better pointed than he could possibly do: therefore being aware of this, and that in moving over 300 paces of level ground he could receive no more than five discharges from each individual opposed to him, and perhaps. not so many, he would resolutely advance; he would calculate on the impression his determined approach, in spight of their fire, might make in his favor; and that in proportion as he thus advanced, at a regulated but rapid pace, that the fire of the enemy would become more confused, rambling, and less executive, and that they would not probably wait the shock of troops approaching in this resolute manner,

but

but would rather place their safety in, a
precipitate retreat.

In regard to the commander of the
posted battalion, were he posted ever so
advantageously, yet if he found his fire
prove as little effective as has been supposed in the first instance, and that by the
manner of the enemies approach he saw
they were determined on the attack sword
in hand, he would, perhaps, were his men
to be depended on in point of courage
and discipline, not wait the attack, but
boldly move forward to meet them, and
thus not only disconcert the enemy in the
midst of his intended plan, by this unex-
pected movement, but in his turn become
the attacking party, (5) and possibly throw
the other on his defence, and by approaching to close quarters dispute the victory
as brave men should, with the naked steel
(6); a circumstance that however unac-
countable it may appear but seldom occurs, and is perhaps to be lamented, for if
troops moved to the charge on both sides
in

in good earnest, the battle would be over in a few minutes, were the bayonet alone to decide it (7).

REMARK (K).

(6th Experiment.)

Charge of Infantry against Infantry in Line.

IN Experiment No. 5, as the attack of infantry in line was represented by the movement of a single man only, and might therefore be supposed to partake more of the nature of a charge *à la débandade* than of a body of men in line, and preserving their dress, it was desirable therefore to try it with a greater number, to prove in what time they could, in dress,

and

and at the regulated step, go over 300 paces of ground.

In the last Experiment the ground was perfectly level and dry under foot: in this, it was uneven all the way, the hollows full of high rushes, and wet; at Flag No. 4 the officers and men were nearly mid-leg in water (1); the men marched over the 250 yards in 145 seconds, preserving the most correct dress, in which time the defending soldier fired seven times. The Experiment was tried a second time, but the result was the same.

This Experiment may shew the necessity there must be for commanders to have their *coup d'œil* correct, in order to be able to guard against the illusions of ground, whether undulating or level, or intersected by obstacles, so as to judge of the distance, and to ascertain the time it would necessarily take troops to traverse such. Thus in the sort of ground here mentioned, the commander of the defending

F   bat-

battalion might know, that if regularly
approached by infantry in line, he could
probably give them seven discharges before they could attain him with their
bayonets; and on the other hand, the
attacking officer might be sure that he
could move over 250 yards of such ground
in 145 seconds, in dress, and at the regulated pace, and that he could in such
time receive possibly seven discharges
from the enemy, but probably not so
many, before he could charge them with
his fixed bayonet.

RE-

REMARK (L).

Infantry in Line attacking Artillery.

IN Experiment No. 2, made in Hyde-Park, the charge of infantry on artillery was represented by a single man ; and for the reasons given in the 6th Experiment, Remark (K), it was desirable to try it with a body of men ; and as Experiment No. 2 was made on dry and level ground, so this was made on a hard sand, at low water, in the presence of Lieutenant-General Gordon, and a number of field officers, &c. &c.

The Experiment, as far as it went, was perfectly complete; for no gun could be better served, nor no infantry better drilled.

F 2　　　　　By

By the trial here made, the leader of a
battalion, or any of its divisions, might
be certain, that with troops, such as were
here employed, he could, in correct dress,
go over 300 paces of level ground in one
minute and fifty-seven seconds, in which
time he might be liable to receive fourteen discharges from a six-pounder, and
no more, probably not so many for reasons
already given; this would enable him to
form his calculations, and make dispositions accordingly, at the same time being
aware, that very quick firing is not so
destructive as might be supposed, by
those who are unacquainted with the effects of artillery, as we are told by an
authority, that on this important subject
must be unquestioned, " that not more
than ten cannon shot in one hundred take
place in an action" (1).

This was the last Experiment made; all
were complete, so far as they went, and it
is hoped may yet lead to matters of great
importance: others, as has already been
observed,

observed, were intended to have been tried in Jersey, but circumstances intervened that prevented their being carried into effect (2).

Junior in rank in the army to many others who no doubt were better informed on these matters, and had better means of being so, yet I had not to complain of meeting many obstructions on that account, nor of giving offence by offering suggestions on subjects of so much importance. " Too many," says Templehoffe, " are inflexibly obstinate, and not only take offence at any suggestion, but frequently endeavour to frustrate the good effects which might result from the ideas of others, merely because they were not their own" (3). So situated, however, it may be readily conceived, that inquiries of the nature attempted here could not be always carried to the desired extent. From the Experiments that have been submitted, and the comments made on them, it is presumed to be evident, that

F 3   the

the intent has been to shew how the service might be benefited by exercising
the different arms together; and that by
thus bringing them acquainted with each
others powers, they might know what each
had to apprehend from the attack of the
other, and the means each possessed in
itself to resist or give the assault; that
they might be sensible, that each had its
weak point where it was vulnerable to the
attack, and that none could justly claim
a superiority over the other; and that
thus they would become convinced of the
necessity of mutual support in the trying
moment of action. In short, by frequently exercising infantry, cavalry, and
artillery, and opposing them to each
other in large bodies, and in every variety
of ground, it might serve in some measure
in place of what is called practice or experience (4) in war; and this, aided by
theory (for they must go hand in hand)
(5), would almost be sufficient to form
the soldier.

Wherever

Wherever it could be done with conve-
nience, each arm might be made to attend
at times to the exercise and manœuvres of
the other, on their several fields of exercise; thus each would obtain an idea of
the tactic of the other, and this might be
further aided by the officer instructing the
soldier, and answering, with mildness,
questions pertinently and modestly proposed (6).

In making the several Experiments,
matters of some importance were ascer-
tained, that before were vague, undeter-
mined, or not generally known. The different military powers of Europe seem to
attach more value to the fire of the infantry than it will be found to deserve,
except given on the *pied firme*, that is, on
the halt; or where soldiers are posted,
or that it is made use of in defence.
British troops possess as much fiery courage and impetuosity as the French (7);
more firmness, and a determined resolution that will not shrink from the edge of

F 4   the

the sword: add to this, that they are, generally speaking, stronger men. We claim
the bayonet as our national weapon (8):
why then should we not pay as much at-
tion to the use of it as to that of the
fire? The inefficacy of the third rank
of firelocks has been shewn in a former
work on this subject (9) as well as here;
and a rank of pikes has been recommended in their stead, for reasons that have
been given, as better adapted to British courage, and the bodily strength of
the men; and it will be recollected, that
the pike held its place in our infantry even
when they were as expert at the firelock
as they are at this day; for it was not (as
we learn from the best authority) laid
aside in our armies, " until (to make use
of our author's words) between the years
1690 and 1705" (10).

Cavalry as well as artillery are frequently exercised where soldiers might
have opportunities of witnessing the manœuvres of the one, or the effect of the
ranges

ranges of the shot of the others: and while interesting trials of this latter description have been making, the soldier has been prevented from availing himself of this opportunity of improvement, by the tiresome routine of perhaps an useless parade, where he is worn down with long standing in a constrained and unnatural attitude (11). In none of the armies of Europe is the exercise of the infantry soldier well calculated to make him active, or capable of undergoing much fatigue; the firelock exercise affords but little scope for muscular exertion; and except while at the drill, or pacing the unvarying round of a given set of manœuvres, the employment of the soldier is sedentary. A large portion of the infantry are taken from men whose occupations are sedentary in the first instance; but it is admitted by the military of all times, that the best soldier is he who is taken from the tail of the plough, the brawny peasant; because, as Vegetius observes, such men are " inured to labour, to carry burdens, to sink ditches,

ditches, to fell timber," in short, to all manner of hard work in the open fields (12).

> Behold the labourer of the glebe, who toils
> In dust, in rain, in cold and sultry skies,
> Robust with labour, and by custom steel'd
> To every casualty of varied life;
> Serene he bears the peevish eastern blast,
> And uninfected breathes the mortal south.
>
> *The Art of Preserving Health,*
> *by J. Armstrong, M.D. Book 3.*

To bring the soldier by degrees to this state of bodily vigour, he should at times be made to handle the hatchet, the mattock, and the spade, taught to throw up works on commons, and level them again (13); used to take long walks into the country, and in marching order; and to form line of battle in the most intersected parts, mountain, hill, or valley (14); to march in line 1500 or 2000 paces over ground rugged or plain, at the quick time, in dress and without a halt; a day in every week perhaps should be allotted for relaxation, for sports, and recreations, (15); small prizes should be allotted to him who fired best at the target, for

excelling

excelling in athletic exercises, wrestling,
swimming. (16), or the race; encourage
them to pursue their national rural games,
cricket, hurling, golf, &c. and while the
sober and attentive soldier is allowed
these indulgencies, the drunkard, the dirty,
the sloven, and the idle should be ex-
cluded, and chained to the utmost se-
verity of the drill. To a soldier thus habituated to muscular exertion, a forced
march would have no terrors; a field day
would be an amusement to him (on the
principles already mentioned, of having
the intention of each manœuvre explained), thus would he conceive a relish for
his profession, become active and vigorous, strong in arm, swift of foot, and
hardy in constitution. The ranks would
be full, the hospitals empty.

———————— By arts like these,
Laconia nurs'd of old her hardy sons;
And Rome's unconquer'd legions urged their way,
Unhurt, thro' every toil in every clime.

ARMSTRONG, *Book 3, line 35.*

NOTES

AND

AUTHORITIES.

N O T E S,

&c. &c.

N O T E (1).

" THUS," says he, " if a body of cavalry beginning the charge has 600 paces to make, it should first advance on the small trot 200 paces, then 200 more on the full trot; this progressive motion, when the horses begin to warm and get wind in a gradual manner, would almost acquire acceleration of its own accord: the remaining 200 paces should be made on a gallop, and when approached to the last fifty, the men should slacken their reins, and put the horses on their full career, so that the greatest velocity could continue to that body they direct their charge upon."—*General Essay on Tactics,* vol. 1, p. 269; also see p. 296.

Marshal

Marshal Saxe, who wrote before M. Guibert, speaking of the charge of cavalry, says: " In charging, they are first to move off at a gentle trot to the distance of about 100 paces; from thence to increase their speed in proportion as they advance, till they fall at last into a gallop; but they must not close to the croup, till they come within about twenty or thirty paces of the enemy; and even then they are to receive the following word of command, as a signal for it, from an officer, *follow me!* This manœuvre is to be performed with the utmost celerity; they must therefore be familiarized to it by constant exercise."—*Reveries or Memoirs upon the Art of War, by Field Marshal Count Saxe,* 4to. p. 56.

NOTE (2).

" The charge," says M. Guibert, " is that motion which the great King of Prussia used to call the *charge in career,* and which he performed with several regiments, without a squadron deviating from the line." To which is added, in a note, " That the Austrian and Prussian

Prussian cavalry charge, *ventre à terre* (on full
speed), and perform their wheels and most of
their interesting manœuvres in the same rapid
manner, for which reason many of their men are
rode over by their horses; but this is never
looked upon as a matter of moment in such
great armies.—*Gen. Ess. on Tac.* vol. i. p. 269.

A Prussian officer remarked to the writer of
this, that such was the strictness of the discipline introduced by the great Frederick into
their armies, that if a cavalry man was thrown
in manœuvring, or if his horse fell under him,
he was sure of being punished in either case, as
it was supposed that neither could have happened without some neglect of the rider.

NOTE (3).

Thirty paces, as mentioned by Marſhal Saxe,
fifty yards, by M. Guibert, or even eighty, as
laid down in our Cavalry Regulations, S. 11,
p, 32, as the proper distance for commencing
the full career, seems to be a very trifling one
indeed to horses that are in any wind. Marshal

G   Saxe

Saxe observes, " that it is above all things necessary, that cavalry should practice gallopping
large distances; a squadron that cannot charge
two thousand paces at full speed without breaking, is unfit for service; it is the fundamental
point, for after they have once been brought to
that degree of perfection they will be capable of
any thing, and every other part of their duty
will appear easy to them."—*Reveries*, p. 56.

In another place the same great man observes,
that the charge of cavalry ought to be as rapid
as the flight of a bird; his words are " sa charge
doit être aussi rapide que le vol d' un oiseau."
*Esprit des loix de la Tactique, ou Notes de M.
le Mareschal De Saxe, commentées par M. M.
De Bonneville, tome premier*, p. 144. General
Lloyd says in his admirable preface to the History of the War in Germany, " that cavalry
should proceed at once to a gallop." The Prussian cavalry are formed upon this principle.

Marshal Saxe's idea of training cavalry to this
point deserves consideration. Speaking of the
dragoons, he says, " in time of peace, and in
winter quarters in time of war, their horses are
to be violently exercised, at least three times a
week, in order to inure them to fatigue, and to
keep

keep them in wind: the same severe usage is also proper for the heavy cavalry at those times; for they must never be spared, or tenderly treated, but in the field where they are constantly exposed to hardships."—*Reveries*, p. 56.

This system is pursued by the French. See a small pamphlet intitled " *Remarks on the Inefficacy of the Third Rank of Firelocks, &c. &c.*" where, in a note, a regiment of French dragoons is described at exercise, near Paris, in 1802.

Marshal Saxe was no advocate for fat horfes: he would have the cavalry mounted on horses, as he says, " inured to fatigue; and above all, that common error of making the horses fat should be avoided: that immoderate love which we are apt to have for the horses, leaves us ignorant of their real power and importance. I had," says the marshal, " a regiment of German horse in Poland, with which I marched, in eighteen months, above fifteen hundred leagues; and I can also affirm, that at the end of that time it was fitter for service than another whose horses were too full of flesh. Unless cavalry be able to endure fatigue, they are, in reality, good for nothing; but then they must be broke to it by degrees, and familiarized to it in length of time by

G 2   custom;

custom; after which, galloping at full speed by squadron, and a constant use of violent exercises, will both preserve them in better condition and make them last much longer: it will moreover form the men, and give them a martial and becoming air."—*Reveries*, p. 42.

N O T E (4).

" The word *movement* seems to be uncertainly applied by the military, many adopt it as consonant with *manœuvre*, others the same as motion, and some think it an awkward term in any sense. There are who conceive, when any great operation is made by a body of troops, *movement*, or *manœuvre*, can be properly applied to it, motion fignifying the small details of the exercise, such as *motions* in the manual, &c."—*Note on Guibert*, vol. i. p. 280.

N O T E (5).

The ancients as well as the moderns, have entertained different opinions, at different times, relative

relative to the best method of drawing up troops. Folard* says, " in the time of Francis the First was one method, in Henry the Third another, another in Henry the Fourth, all different from those practised in his time.† Henry the Fourth, Prince Maurice, and Gustavus Adolphus, as well as other great captains, drew up their infantry ten or twelve in file; in the time of Louis the Thirteenth, eight in file;"‡ and §Puysegur tells us " that eight in file was the usual method in the minority of Louis the Fourteenth, though the armies were then small. The Prince of Condé and Marshal Turenne did the same at that time. However this number was soon changed; for the first of the three last wars, he says, they formed the battalions only six in depth; the second, five; and the last, only four, and then three:" the reason was, upon the suppression of pikes to be able to command more fire. The above is from the *Introduction to the Target*, p. 27.

The first who departed from the order of depth was the great Gustavus Adolphus: he

* Folard, tom. iv.
† Tom. i. Traite de la Colonne, de Folard.
‡ Préface, tom. iii. de Folard.
§ Puysegur, ch. v. art. 1.

drew

drew up his men six deep, while Tilly and Walstein had theirs in solid masses thirty deep."
*Military Miscellany, by the Honourable Colin
Lindsey.* "Gustavus was looked upon as a bold
innovator, but success justified this truly great
man; he first showed in the north of Europe,"
says M. Guibert, "the phenomenon of a warlike
and well disciplined army, created by himself.
Under him, and the great Nassau, (Maurice,
Prince of Orange) military science once more
took its birth. On the principles laid down by
Gustavus—Bannier, Gassion, Weimar, Turenne,
and Montecuculi, fought. They continued the
custom of the pike; they still believed the force
of the infantry consisted in the density of its order, and its impulsion." See the enlightened
and elegant *Preliminary Discourse of M. Guibert, General Essay on Tactics*, p. xlv,

---

NOTE (6).

"In Turenne's days" (from 1624 to 1675),
"troops were ranged eight deep, both in France
and Germany; thirty years after, in the time of
Puysegur, the ranks were reduced to five, in the
last

last Flanders war" (1756) " to four, and immediately after to three."— *Elementary Principles
of Tactics, &c. by Sieur B——, Knight of the
Military Order of St. Lewis,* p. 136, a valuable
work, and worthy the attention of every officer.
In the times above alluded to the cavalry were
formed eight deep, then six, now they are universally formed only two.

NOTE (7).

" Those great men, Gustavus Adolphus, and
the Prince of Orange, greatly resembled the ancients in having small armies, which made the
attirail of their camp equipages few in number,
&c. This system of the military was in many
respects kept up in the time of Turenne. This
great man gave the preference to the commanding of small armies. Louis XIV. was the first
who set the example of having numerous armies.
The only thing he did was the engaging Europe
to imitate his principles. When it is difficult
to maintain and put armies in motion, the commanding of them must, of course, be likewise
perplexing and difficult; superior genius alone
knew

knew how to move those unwieldy masses. The great King of Prussia showed Europe the phenomenon of an amazing army, manœuvring, and well disciplined: he divulged this secret to the world, that the movements of one hundred thousand men are subject to as simple, and as certain calculations, as one thousand: that the spring which puts one battalion in motion, being once found, the question then only remains, how to combine a greater number of those springs, and afterwards how to work them. His victories have been sufficient proofs of the justness of his discoveries." — See *General Essay on Tactics, Preliminary Discourse,* p. xlvi. also p. 98, and 300, of the Essay, vol. i. In giving a preference to small armies, Gustavus had, no doubt, the Roman model in view, for " a consular army seldom consisted of more than 50,000 men; at the battle of Canna the armies of the two consuls amounted to 87,000 men."—*Memoirs de Montecuculi,* l. ii. c. ii. p. 269.

NOTE

NOTE (8).

The horsemen alluded to, were denominated at different times, " cataphractes lanciers,"* or lance men. " Gens armés de toutes pieces, that is men in complete armour."—*M. de Montecuculi,* p. 11, 14. " Gendarmerie."—*Gen. Ess. Tac.* vol. i. p. 262. " Men at arms."— *Grose's History of the British Army.*

" The lancier, lance man, or man at arms, was in complete armour: he wore a close helmet, with a vizor to lift up and down, or one with a vizor and bever† both revolving on the same pivot; the vizor was opened to obtain a less obstructed sight, and the bever to enable the wearer to converse more freely, or to eat and

---

* " The cataphractes, or men armed cap-a-pie, gens armés de toutes pieces, were no more to be seen in 1669." *Montecuculi,* l. i. p. 11. " L'armure de pied en cap, was abolished in France in the reign of Henry IV.—*Ecole de Mars,* tom. ii. l. v. p. 3.

† " Helmets with bevers, were not in use till the middle of the fourteenth century."—*Grose's Hist. Brit. Army,* vol. i. p. 104.

drink.

drink. When these were closed, the air was admitted through apertures made also for sight, and other smaller perforations for the mouth and nostrils. The neck and breast was defended by a gorget, or hallercet; the body by a cuirass‡ or

‡ " There was what was called the intire cuirass, les cuirasses entières avec le devant et le derriere."—*Montecuculi*, as above. " The breast plate (or fore part of the cuirass) was one piece of solid iron that reached from the man's chin to the saddle bow, which was made of steel, and reached as high as the man's navel. The hauberk was the knight's armour, it consisted of the back and breast plate of solid iron; but the hauberk was changed, as was the solid iron cuirass, for the plated armour, that came into general use about the middle of the fourteenth century: the plated armour was made of plates of iron that were pliable with the motions of the body."—*Grose's Hist. of the Brit. Army*, vol. i. p. 104.

" When fire arms were first introduced ' the knights,' says Le Nou, ' that their armour might be musket proof, loaded themselves with anvils, instead of covering themselves with armour.'—*Grose's Hist. of the Brit. Army*. Some idea of the strength of their armour may be conceived by the following curious anecdote.

" At the battle of Tournois, a number of Italian knights who were overthrown, could not be slain on account of the strength of their armour, till broke up, like huge lobsters, by the servants and followers of the army, with large wood cutters axes, each man-at-arms having three or four men employed about him."—See *Grose's Hist. of the Brit. Army*, vol.

(or corslet), formed of two pieces hooked together, denominated backs and breast plates; to
the

vol. i. p. 105. Le Nou, from whom Captain Grose has taken
the above, was a famous French commander, he was a Hugo-
not, and served under Charles IX. Henry III. and Henry IV.
he was surnamed Iron-arm, and was universally esteemed for
his probity and for his valour.

" La Noue, fameux capitain Francoise, Hugenot, sur-
nommè bras de fer, il a servi sous les Rois Charles IX. Henry III. Henry IV. il a éte généralement estimé pour sa pro-
bité et pour sa valeur."—*Memoires De Montecuculi*, l. ii. c. xi.
p. 269.

" As fire arms came more in use, the lance, though styled
by Montecuculi the queen of arms for the cavalry, as the pike
for the infantry, (l. i. c. ii. p. 17.) was yet liable to many ob-
jections; it was heavy out of action, and unwieldy in it; covered almost half its length, from the blade downwards, with
plates of iron, to preserve it from the sword of the enemy, it
required that the cavalier who wielded it should be vigorous
and robust, the horse he rode strong, and highly dressed, and
the ground he charged over free from impediments, dry and
firm under foot, and unless all these circumstances corre-
sponded the lance often remained a useless weapon in the hands
of the horseman; it was therefore laid aside, and with it the
musket proof armour, which was found to be intolerably
heavy, as well for the man as the horse, and so embarrassing
that if the horse fell, or the cavalier was otherwise unhorsed,
he lay immoveable on the spot, unless assisted by his friends,

or

the back was joined a *gard de reins,* or culet; the arms were covered with brassarts, called also avant-bras, and, corruptly, vambraces; the hands by gantlets; the shoulders by poul-drons; the thighs by cuissarts; and the legs by iron boots, called greaves, and sometimes by boots of jacked leather (cuir bouillée); over all these was worn a jacket of thick fustian, or buff leather. When these arms were well tempered they were proof against the lance or the sword, until the use of gunpowder, and then the casque was made musket proof, as was the fore part of the cuirass; the back part of it was made pistol proof."—See for the above *Grose's History of the British Army.* " They were armed with lances and long straight swords."—*Montecuculi,* l. i. c. ii. p. 17. " They had also for offensive arms, which they frequently carried, iron maces

or squire (valet), to remount, or get him off the field, for unless this was the case he was at the mercy of the enemy. However they generally endeavoured to unhorse the knights that they might take them prisoners, in order to have them ransomed, otherwise they dispatched them with a small dagger called a *miserecorde*; it was small, otherwise it could not find way through the joining of the armour: so close was the armour that in heat and dust the knights were often smothered in it."—*Grose's Hist. of the Brit. Army,* vol. i. p. 103.

suspended

suspended at their saddle bow."—*Grose's Hist. Brit. Army*, vol. i. p. 103.

The horses also were covered with complete armour, as well as the cavaliers ; the head, and all the body was covered with, and caparisoned in iron. Les chevaux étoient anciennement armés de toutes pieces, comme les cavaliers ; la tête & tout le corps étoient couvert & caparaçonnes de fer."—*St. Remy Mem. d'Artillerie*, *tome premier*, p. 298.

" The faces and ears of the horses were covered with a sort of mask, so contrived as to prevent their seeing right before them, in order that they might not be terrified from charging, or shocking, with vigour. This mask was called a *chafron*, or shafront.* Frequently from the centre of the forehead projected an iron spike : their necks were defended by a number of small plates of iron connected together, called a criniere, or manefaire; they had

* Armour for the horse's head was called a *chamfrain*.

" Armure de fer pour tête de cheval : cette sort de armure se nomme chamfraine."

Mem. d'Artil. de St. Remy, tom. 2. p.106.

poitrinals

poitrinals for the breasts; cropieres and flaneois for covering their buttocks, reaching down to the hock : all these pieces were generally of iron or brass, though sometimes of *cuir-bouillée,* i. e. jacked-leather. Occasionally they were covered all over with mail, or linen stuffed, and quilted like the gambeson, and adorned with rich embroidery; they had plumes of feathers on their heads, and on their rumps. Horses thus covered were called *barded,* and, corruptly, *barbed** horses. They were so loaded with men and

* *Bardè,* in old French, according to Grose (*Hist. British Army*), signifies covered.

Barbed horses were well known formerly in England; so in Richard the Third :—

 " And now, instead of mounting *barbed* steeds
 " To fright the souls of fearful adversaries"—

According to the commentator on Shakspeare :—
 " Steeds caparisoned in a warlike manner."

Barbed, however, may be no more than a corruptioh of *barded.* Equus *bardatus,* in the Latin of the middle ages, was a horse adorned with military trappings. Barded occurs many times in our old English chronicles; an instance or two may suffice:—

They mounted him surely upon a good and mighty courser, well *barded.* *Knight of the Swan.*

Again

and arms, that to preserve their vigour for the
charge, the men at arms had commonly hackneys,

Again, in Barrett's *Alvearia*, 1580:—*Bardes* or trappers
of horses.—We learn that *bardes* and trappers had the same
meaning.—*Shakespeare's Richard III. Note 4, by Stephens,*
vol. x. p. 461.

The armour of these horsemen was of iron or steel highly
polished; and when well tempered, was proof against the
lance, or swords of any description; and when mounted
on horses beplumed and covered with shining armour of
proof, must have had a grand and imposing effect. "Leur
habillement estoit de fer, ou acier, bien luisant, bien poli, &
bien trempe. Quand ces armes estoient bien trempees, elles
garentissoient du coup de lance, du coup d'épée, du coutelas,
ou du sabre."—*Mem. d'Artillerie,* part ii. p. 297.

" No one," says M. Guibert, " can view history without
feeling, in some degree, for the great ignorance of former ages.
The *gendarmerie,* like a ponderous body of iron in motion,
charged on a much slower pace than a trot: if the ground had
imbibed any wet, they had no ability to proceed; thus they
were perishing under the pressure of their useless armour, and
annoyed either by the archers, or by the attack of a much
lighter cavalry."—*G. E. on Tac.* vol. i. p. 264.

However, when all circumstances favoured, their shock
was not to be resisted; every thing gave way before it; and
where the armour was proof against the missive weapon of the
enemy, their proceeding *au petit pas* was an advantage ; as by
that slow movement they preserved their dress, a matter now
so much insisted on.

It

neys for riding on a march, and did not mount
their war horses till they were certain of coming

.   to

It may be remarked here, that there were as various opinions
about forming the cavalry, as the infantry. They did not
begin to form squadrons in France, till the reign of Henry the
Second. When they used lances, they charged in a single
line; a custom they with difficulty changed. The Prince of
Condé ranged his cavalry in this manner at the battle of St.
Dennis, under Charles the Ninth : this was the method practised by the gens d'arms, when they were in their perfection;
a method those great captains, the Duke of Guise and the
Constable, &c. did not wish to alter; it had its advocates as
being the best method, but not long after it was thought not
to be the best. They formed their squadrons of four and
five hundred horse each, and ten in depth ; but this did not
answer. The battle of Coutras, in 1587, in the reign of Henry
the Third, was lost, in a great measure, by the large squadrons of Joyeuse's lanciers. The young King of Navarre,
and the Princes of Condé and Conti, running in upon them,
rendered them useless. In 1590, Henry the Fourth gained
the victory of Ivry chiefly by the same means; the great
squadrons of lances that were sent to the Duke of Mayene
(who was the head of the Leaguers), by the Duke of Parma,
from the Low Countries, being heavy and unskilful, were
broken to pieces by the others swords and pistol shot. By
degrees, however, they diminished both their number and
depth. Henry IV. of France, Prince Maurice who commanded the Dutch, Alexander Farnese, and the Duke of
Alva, who were at the head of the Spanish armies in the
Low Countries, formed their squadrons eight in depth, and
then six. And afterwards Walstein, the Emperor's general,
made

to action. Sometimes the men at arms were surprized by their enemies on their hackneys, and
destroyed before they could get on their war
horses. They were all stone-horses, it being a
mark of disgrace for a knight to ride on a mare.
Barded horses were in use in England in Edward
the Sixth's time."—*Grose's Hist. B. A.* vol. i.
p. 104.

made his very large and deep; he found the inconveniency
of them at Lutzen, where he was defeated by the Swedes the
6th of November, 1632, a battle dearly purchased by them,
as it cost the life of their king, the great and good Gustavus
Adolphus. Count Tilly found the same inconveniency from
the great depth of his squadrons at Leipsic (bataille de Brei-
tenfeld près de Leipsic), where he was defeated the 7th of
September, 1631, by the great Gustavus. That enlightened
hero was of a different opinion from Walstein and Tilly, his
antagonists: he was the first (as has been mentioned before)
who departed from the order of depth; he formed his cavalry
in small squadrons, and five in depth, with platoons in the intervals. Marshal Turenne ranged his four, and sometimes
five in depth, as did his antagonist Montecuculi. In the
war 1701, most of the cavalry in Europe was formed three
deep, and about 150 in a squadron. " Thus," says the author of the *Target:* " What one age does, another undoes;
what one approves, another condemns." Whoever would
wish to see more on this subject, may consult the *Target,*
pages 26, 27, and 28; also the authorities he quotes: *P. Dan.
Hist. de la Milise Franc.* tom. i. liv. 5, p. 314—*Mezeray—Folard,* tom. iv. 1. 3, ch. 13; and *Le Nou—Puysegur,* tom. i. ch. 5
—*The Chevalier Ramsey's Life of Turenne*—also *Montecuculi*—
and *M. Guibert,* as already quoted.

H   NOTE

N O T E (10).

" By the greatest degree of speed of cavalry is not meant (says M. Guibert) the greatest degree of a single horseman, left entirely to his own discretion, but the greatest degree of velocity of a troop, keeping at the same time its order: and in proportioning this velocity from the spot it departs from, to the spot it is carried to, and to the object which it should accomplish at its arrival."—*G. Essay of Tactics,* vol. i. p. 263.

N O T E (11).

M. Guibert observes, that " when the most perfect dress of the line is mentioned, there is no necessity to require the minute attention of neither squadron or horse swerving from it, there being only wanted a sufficient concert of the whole of the squadrons ; so that the line may join the enemy nearly at the same time on all points."—*Gen. Essay of Tactics,* vol. i. p. 295.

Notwith-

Nothwithstanding the value attached to the charge in line, and in dress, General Lloyd appears to have been of opinion, that the charge *à la débandade*, so much condemned by M. Guibert, and so universally exploded now, was nevertheless the most executive, as " one man (says he) acting in this manner has more real activity than seventy who advance and attack in a line, as usual. I once saw three hundred horse attack a column of seven or eight thousand foot in this way, which they defeated, and dispersed, in three or four minutes."— *Lloyd's Rhapsody*, p. 63.

NOTE

N O T E (1).

(Remark B.)

" THE paces of cavalry cannot be regulated by length of step, and numbers in a minute, as those of the infantry, nor is it so material."— *Cav. Reg.* §. ii. p. 29.

N O T E (2).

" The walk, trot, and gallop, are the three natural paces of the horse, and of each of these there are different degrees of speed."—*Cavalry Regulations.*

These different degrees of speed are perhaps what make it difficult to prove the paces of a horse, but do not render it at all impossible. Colonel Herries, who it appears paid attention to this subject, mentions: " That horses generally step three feet at a walk, and four at a trot; a

horse

horse that walks at a regular extended pace will go four miles an hour, that is, a mile in fifteen minutes; and taking as he steps about a yard he takes nearly 120 paces in a minute; when he trots, he steps about four feet every time, and to trot twelve miles he must take 264 steps, or the fifth part of a mile every minute."—*Light Horse Drill, by Col. Herries.*

In the walk, the horse (as above mentioned) takes 117⅓ paces in a minute, which is nearly the wheeling time of the infantry.

The *Bombardier* and *Pocket Gunner* observes, " That military horses walk about 400 yards in four minutes and a half; trot the same distance in two minutes and three seconds; and gallop it in about one minute."

According to Colonel Herries, the horse moves at the following rates:—In his walk he takes three feet at each step, or 117¾ yards in a minute, or four miles an hour; trots four feet at each step, or 350 yards in the minute, or twelve miles an hour.

According to the *Bombardier*, the walk of the horse is at the rate of three miles and 53¾

H 3   yards

yards an hour; the trot nearly seven miles an hour; and the gallop nearly fourteen miles an hour.

It appears in the experiment made here, the dragoon horse walked at the rate of one mile in thirteen minutes and a half, or about four miles and a half in an hour; trotted twelve miles an hour nearly; galloped and charged one mile in 130 seconds. All his paces taken together, he moved over the 600 yards in 144 seconds, or one mile in seven minutes and a half, nearly.

It may be remarked, that according to the work above quoted, " a light dragoon horse mounted and accoutred complete, carries about 2 cwt. 1 qr. 14lb. without forage; a day's forage would be about twenty pounds more." See the *Bombardier* and *Pocket Gunner, by R. W. Adye, Captain of the Royal Regiment of Artillery*, second edition.

Colonel Herries, in his *Light Horse Drill*, already quoted, recommends, " To prove the paces of the slowest moving horse, and to place such horse on the flank to regulate the movement of the squadron." In fact, the movement

ment must (according to our tactic) be regulated
by the paces of the slowest horse.

" At whatever pace the squadron is conducted,
the slowest moving horse at that pace must be
attended to, otherwise different kinds of motion
will exist at the same time, and tend to dis-
unite it." (See *Cavalry Regulations*, §. ii. p. 32);
and the reason is evident, for it is in the uniform
velocity of the squadron that its effect consists,
as inequality of step would lessen much the effect of the charge; because the line would not
be in dress, and consequently could not join the
enemy at the same time on all its points."—See
Cav. Reg. as above.

NOTE (4).

" When engineers," says Captain Tielk, " allow their guides to pace the distances, it is
highly probable that their paces will be incor-
rect; therefore as they cannot measure the
whole on foot, they may procure steady horses
and abide by their paces, for they are always

more

more even than those of men."—*Tielk's Field
Engineer*, vol. i. c. vi. p. 209. *Colonel Hewgill's Translation*.

"An officer," says the author of the Light
Horse Drill, already quoted, "may calculate
the extent of his horse's walk and trot, this will
enable him to take up ground and form his calculations accordingly." And Captain Tielk observes, "as some horses make longer paces than
others, it is of course necessary that every officer should know how many his own will take."
See *Tielk*, as above, vol. i. p. 209.

NOTE (5).

From what has been said by some of the most
distinguished writers on the art of war, it would
appear that among the whole circle of military
knowledge nothing is so essential to a commander as the possessing a correct military eye, or
what the French call *coup d'œil militaire, or coup
d'œil à la guerre:* a few extracts on this interesting subject are here given from great authorities,
which it is presumed may not be found intirely
useless

useless to young officers, to point out to them its importance; in what it consists; and how it may be obtained; or at least improved. "It is generally imagined," says M. Folard, "that the *coup d'œil* does not depend on ourselves; that it is a gift from nature; that it is not to be acquired in campaigns; and in short that it must be born with us, without which the most piercing eyes in the world can see nothing, and we walk in the thickest darkness; but," continues the chevalier, "they deceive themselves, we have all the *coup d'œil*, according to the degree of genius or good sense that it has pleased Providence to bestow on us; it is born with us; knowledge refines and brings it to perfection, and experience confirms it to us; his words are, "c'est le sentiment général que *le coup d'œil* ne dépend pas de nous, que c'est un présent de la nature, que les campagnes ne le donnent point, et qu'en un mot il faut l'apporter en naissant, sans quoi les yeux du monde les plus perçans ne voyent goute, et marchant dans les ténébres les plus épaises. On se trompe; nous avons tous le coup d'œil, selon la portion d'esprit et de bon sens qu'il a plû à la Providence de nous départir. Il nait de l'un et de l'autre, mais l'acquis l'affine et le perfectionne, et l'experience nous l'assure." He then goes on to define it: "the

military *coup d'œil*," says he, " is nothing else
but the art of understanding the nature and different situations of the country which is the
seat of war, or into which we would wish to
carry it; the advantages and disadvantages of
the camps and posts that we would wish to occupy, as also of those which might be advantageous or disadvantageous to the enemy. From
the position of our troops, and the consequences
we derive from it, we can judge as to our present plans, and what we may propose for the
future. It is only by this knowledge of the
nature of the whole of a country, in which we
carry on war, that a great 'general can foresee
the events of an intire campaign, and make
himself in a manner the master and disposer of
them; for judging by what he does himself of
what the enemy must necessarily do, who is
obliged, by the nature of the country, to regulate his movement according to his plans, he
conducts him from camp to camp, and from
post to post, to that point which he has fixed
upon in his own mind to gain the victory.
This is, in short, what is meant by the military
coup d'œil, without which it is impossible for
a general to avoid falling into a number of
faults of the greatest consequence; in a word,
there is no hope of victory if one is not possessed

sessed of this *coup d'œil*: the words are, " le coup d'œil militaire n'est autre chose que l'art de connoître la nature et les différentes situations du pays où l'on fait et où l'on veut porter la guerre, les avantages et les désavantages des camps et des postes que l'on veut occuper, comme ceux qui peuvent être favorables ou désavantageux à l'ennemi. Par la position des nôtres, et par les conséquences que nous en tirons, nous jugeons surement des desseins présens, et de ceux que nous pouvons avoir par la suit. C'est uniquement par cette connoissance de tout un pays où l'on porte la guerre, qu'un grand capitaine peut prévoir les événemens de toute une campagne et s'en rendre pour ainsi dire le maître; car jugeant par ce qu'il fait de de ce que l'ennemi doit necessairement faire, obligé qu'il est par la nature des lieux à se régler sur ses movemens pour s'opposer à ses desseins, il le conduit ainsi de camp en camp, et de posto en poste, au but qu'il s'est proposé pour vaincre. Voilà en peu de termes ce que c'est que le coup d'œil militaire, sans lequel il est impossible qu'un général puisse éviter de tomber dans une infinité de fautes d'une extréme conséquence; en un mot, il n'y a rien à espérer pour la victoire, si l'on est depourvu de ce que l'on appelle *coup d'œil à la guerre.*"

This

This subject may be seen satisfactorily and elegantly treated in the first thirty-eight pages of L'Esprit du Folard, under the head *Coup d'Œil Militaire*, or the art of knowing the nature and different situations of a country where one would carry on a war, the advantages of camps and of posts which one would occupy. That the *coup d'œil* produces the great and admirable in war, that it may be acquired by study and application, and the error of those who maintain that it is a gift of nature; his words are:—" *Coup d'œil militaire*, ou l'art de connoître la nature et les differentes situations du pays où l'on veut porter la guerre, les avantages des camps et des postes que l'on veut occuper. Que le *coup d'œil* produit le grand et le beau d'une guerre, qu'il peut s'acquérir par l'étude et l'application. Erreur de ceux qui prétendent que c'est un present de la nature."—*Folard*, tom. i. p. 219, à 228, inclufivement. Speaking of Philopœmen, one of the greatest captains of Greece, Folard says that he had an admirable coup d'œil, and that in him one ought not to consider it as a gift of nature, but as the fruit of study and application, and of his extreme passion for war. " Philopœmen, un des plus grands capitaines de la Grèce, qu'un illustre Romain appella le dernier des Grecs, avoit un *coup d'œil* admirable: on ne doit pas

le

le considérer en lui comme un présent de la nature, mais comme le fruit de l'étude, de l'application, et son extrème passion pour la guerre."
p. 6.---His method (says Folard) was, " when he
had read the precepts and the rules of tactics,
he did not think it necessary to see the demon-
strations of them by plans upon paper, but he
made the application of them on the ground
itself, in the midst of a campaign; for in the
marching, he observed exactly the position of
hills and vallies, all the intersections and irregu-
larities of ground, and all the forms and figures
that battalions and squadrons are obliged to
take, on account of rivulets, ravines, and defiles,
which force them to diminish or extend themselves; and after having meditated upon all these
things within himself, he communicated with those
who accompanied him. His words are, " Quand
il avoit lu les préceptes et les régles des tactiques,
il ne faisoit nul cas d'en voir les démoustrations
par des plans sur des planches, mais il en faisoit
l'application sur les lieux mêmes, et en pleine
campagne. Car dans les marches il observoit
exactement la position des lieux hauts et des
lieux bas, toutes les coupures et les irrégularités
du terrein, et toutes les différentes formes et
figures que les battaillons et escadrons sont obli-
gés de subir, à cause des ruisseaux, des ravines,

et

et des défiles qui les forcent de se resserrer ou de s'étendre; et après avoir médité sur cela en lui-même, il en communiquoit avec ceux qui l'accompagnoient." --- See *L'Esprit du Chevalier Folard, tiré de ses Commentaires sur l'Histoire de Polybe pour l'Usage d'un Officier, De Main de Maître*, p. 7. It is perhaps unnecessary to mention that the above valuable work (of which there is no English translation) is in one volume large octavo, 302 pages, with plates, and that it is an abridgment of six large quarto volumes of Folard's Commentaries on Polybius, by the great King of Prussia, who, at the conclusion of his preface, mentions that those who undertook the care of having this abridgment printed, proposed nothing to themselves but the utmost glory of the service, by endeavouring to facilitate to officers the study of their profession, a profession that leads to immortality. His words are, " Ceux qui ont eu soin de faire imprimer cet abrégé ne se sont proposés que la plus grande gloire du service, en tachant de faciliter aux officiers l'étude de leur art et d'un métier qui mène à l'immortalité."--*Avant-Propos*, p. viii.

It may not perhaps be improper to remark here, that M. Folard has drawn the most valuable

able part of his observations on the *Coup d'œil*
from the rich sources of Machiavel, and, what is
not common among military writers, (who are
not fond of acknowledging their obligations to
those authors whom they have not been ashamed
to plunder), he generously pronounces an eulo-
gium on that great and enlightened author;
saying among other matters in his praise, " that
his political and military discourses on the Decades of T. Livy are an immortal work ; an admirable treatise, worthy the attention of the
military, and which should be well read and
considered by them:" his words are, " Les dis-
cours politiques et militaires de cet auteur, sur
les Décades de Tite Live, sont un ouvrage im-
mortel; je le trouve digne de la curiosité des
gens de guerre, et d'en être bien lu et bien
médité."---P. 18. His life also of Castrucio, one
of the greatest captains of his age, although but
little known, is not the less admirable; it is re-
plete with facts that are curious, highly instruc-
tive, and is full of military reflections and observations, that few people are capable of making;
so strongly was his genius turned to military
affairs.--Folard's words are, " Sa vie de Castrucio, un des plus grands capitaines de son
siècle, quoique peu connu, n'est pas moins admirable;

mirable; elle est tout ornée de faits curieux,
très instructifs, et pleins de réflexions et d'observations militaires que peu de gens savent
faire: tant cet homme avoit le génie tourné au
métier."

Besides the admirable discourses of Machiavel
on the Decades of Livy, his life of that famous
commander Castruccia Castricani, of Lucca,
(recommended so strongly, by so great a judge
of military affairs as the Chevalier Folard, as
worthy the attention of all who make arms their
profession), it need scarcely be added that his
admirable treatise on the Art of War, will also
be found well worth the study of the young
officer. The works of Machiavel have been now
published about three centuries, and more use
has been made of them by political and military
writers than they have been forward to acknowledge; and as his book is not in every one's
hands, the curious reader may not be perhaps
displeased to see that part of it which the Chevalier Folard has so judiciously selected for his
observations on the *coup d'œil*, as containing in
itself every thing almost that may give a clear
idea not only of what it is, but of the admirable
use that may be made of it in war. Here follows
the chapter.

" *The*

" *The Discourses of Nicholas Machiavel.*

" B. 3.—Chap. 39.—P. 425.

" *A General ought to know the Country, and how to take his Advantage in the Ground.*

" Among the many things that are necessary in a general of an army, the knowledge of coasts and countries is one, and that not only in a general but in an exquisite and more particular way, without which he shall not be able to do any great thing: and because all knowledge requires use and exercise to bring it to perfection, so it is in this knowledge of places; and if it be inquired what use, and what exercise is required in this case, I answer hunting and hawking, and such like recreations; and therefore it is that the heroes which anciently governed the world, were said to be brought up in woods and forests, and accustomed to those kind of exercises: for hunting (besides the acquaintance which it gives you of the country) instructs you in many things that are necessary in war; Xenophon in the Life of Cyrus tells us, that when Cyrus went to invade the King of Armenia, assigning several offices and places to the several parts of his army, he told them that, " Questa, non era altro ch'una di quelle caccie le quali molte volte ha-

I   venano

venano fatte seco: That this expedition was no more than one of those chaces which they had taken frequently with him." Those whom he placed as scouts upon the mountains, he resembled to them who set their nets upon the hills; and those who were to make excursions upon the plain, were like them who were employed to rouse the deer, and force them into the toils. And this is said by Xenophon to shew the resemblance and similitude betwixt hunting and war: for which cause those kind of exercises are not only honourable, but necessary for great persons; and the rather, because nothing gives a man so true a knowledge of the country, or imprints it more deeply and particularly in the memory; and when a man has acquainted himself thoroughly with one country, he may arrive more easily at the knowledge of another, because all countries and coasts have some kind of proportion and conformity betwixt them: so that the knowledge of the one contributes much to the understanding of the other. But if, before you have acquainted yourself with your own, you seek out new regions, you will hardly, without great labour, and long time, come to the knowledge of either. Whereas, he that is well versed and practised in one, shall, at the first cast of his eye *(coup d'ail)*, give you

an

an account how that plain lies: how that mountain rises; and how far that valley extends;
and all by his former knowledge in that kind.
To confirm all this, Titus Livius gives us an
example in Publius Decius, who being a military tribune in the army which the consul
Cornelius commanded against the Samnites,
and finding the said consul and army fallen,
by accident, into a vale where they might
have been encompassed by the enemy, and cut
off;" " Vides tu Aule Corneli," (said Decius
to the Consul) " cacumen illud supra hostem?
Arx illa est spei salutisque nostræ; si eam (quo-
niam cæci reliquere Samnites) impigre capimus:"
" Do you see, Sir, that hill which hangs over
the enemy's camp? there lies our hope; the
blind Samnites have neglected it, and our safety
depends upon the seizing of it quickly." For,
said Livy, before " Publius Decius tribunus
militum, unum editum in saltu collem, immi-
nentem hostium castris, aditu arduum impe-
dito agmini, expeditis haud difficilem:" Publius
Decius, the military tribune, observed a hill
over the enemy's camp not easily to be ascended
by those who were completely armed, but to
those who were lightly armed, accessible enough;
whereupon being commanded to possess it, by
the consul, with 3000 men, he obeyed his orders,

secured

secured the Roman army; and designing to
march away in the night, and save both himself
and his party, Livy brings him in speaking
these words to some of his comrades: " ite me-
cum, ut dum lucis aliquid superest, quibus locis
hostes præsidia ponant, qua pateant hino cxituo
exploremus, hæc omnia sagulo militari amictus,
ne ducem circuire hostes notarent, perlustravit:"
come along with me, that whilst we have yet
light we may explore where the enemy keeps his
guards, and which way we may make our retreat; and this he did in the habit of a private
soldier, that the enemy might not suspect him
for an officer. He then who considers what has
been said, will find how useful and necessary it
is for a general to be acquainted with the nature
of the country; for had not Decius understood
those things very well, he could not so suddenly
have discerned the advantages of that hill, and
of what importance it would be to the preserva-
tion of the Roman army: neither could he have
judged, at that distance, whether it was accessible or not; and when he had possessed himself
of it, and was to draw off afterwards, and follow
the consul (being so environed by the Samnites),
he could never have found out the best way for
his retreat, nor have guessed so well where the
the enemy kept his guards. So that it must necessarily

cessarily be that Decius had a perfect knowledge of the country, which knowledge made him secure that hill, and the securing of that hill, was the security of the army. After which, by the same knowledge (though he was as it were besieged by the enemy), he found a way to make his own retreat, and bring off his whole party." The above extract is from an English translation, in one volume folio, entitled *The Works of the famous Nicholas Machiavel, Citizen and Secretary of Florence.*" Starkey, 1672.

Speaking of the importance of the *coup d'œil militaire,* Captain Tielk observes, that " all great captains and able engineers have possessed it, and to this they have owed their future reputation and success; in writing on the art of war they have supposed their readers acquainted with this science; as all engineers, when they treat of fortification, suppose a previous knowledge of the mathematics."—*Tielk's Field Engineer, Colonel Hewgill's Translation.*—The " Cadet," a very useful military treatise, defines it thus: " *Coup d'œil,* a quickness in discovering a country proper for encamping, by its situation in regard to plains, mountains, rivers, passes, defiles; security of the camp; conveniency of convoys; covering of our own, or distressing the enemies

I 3   country;

country; and many other circumstances, such as wood, water, forage, &c. &c."—*Cadet*, p. 20.

Monsieur De Jeney, enumerating the qualities requisite for a partizan, mentions the necessity of his being possessed of a correct *coup d'œil*, " A piercing rapid eye, which instantly catches faults or advantages, obstacles and dangers of situation, of country, and every object as it passes." — *English Translation of the Partizan*, p. 47.

The great King of Prussia was thoroughly sensible of its importance. His words are, " the *coup d'œil* may be reduced, properly speaking, to two points, the first of which is, the having abilities to judge how many troops a certain extent of country can contain; this talent can only be acquired by practice, for after having laid out several camps, the eye will gain so exact an idea of space, that you will seldom make any material mistake in your calculations."

" The other, and by far the most material, point, is to be able to distinguish, at first sight, all the advantages of which any given space of ground is capable. This art is to be acquired, and even brought to perfection, though a man
be

be not absolutely born with a military genius. Fortification, as it possesses rules that are applicable to all situations of an army, is undoubtedly the basis and foundation of this *coup d'œil.* Every defile, marsh, hollow way, and even the smallest eminence, will be converted, by a skilful general to some advantage; two hundred different positions may sometimes be taken up in the space of two square leagues, of which an intelligent general knows how to select that which is the most advantageous, &c."—See *Military Instructions from the late King of Prussia to his Generals, translated by Major Forster,* 1797, p. 27, 28.

" Too much attention," says M. Guibert, " cannot be given to the studying of the *coup d'œil,* it is more difficult to acquire in the cavalry than in the infantry, becaufe the motions of infantry are much slower, and where the eye finds more time to measure and compare objects; in a contrary manner, as the movements of cavalry are more rapid, the resolution of an officer should be taken more speedily; and as the points of sight are much more difficult to fix upon, so the least error in the *coup d'œil* is productive of the greatest deviations; in short, the same alacrity with which a false motion is

I 4   made

made, taken advantage of by a skilful enemy, would encourage him to profit by the error."— *Gen. Ess. on Tac.* vol. p. 296. " General Loudon, who commanded the Austrian cavalry at Bergen, near Frankfort on the Main, gained a complete victory over the Prussians by an instantaneous application of the *coup d'œil.* Having found that the artillery had made fome confufion on the enemy's wing, a circumstance apparently invisible to every other eye, he gave the word for the charge, by which he defeated the enemy, and took victory out of the hands of the Austrian allies, the Russians." The above is given as an example; see a *Note to Gen. Ess. on Tac.* vol. i. p. 290. The opinions of such high authority as General Lloyd, cannot be overlooked. " Experience," fays he, " and a certain *coup d'œil* aided by theory, will enable a man to judge of the time and space necessary to execute any evolution whatever; a thing of the utmost consequence in a day of action, because you will be able to make a thousand motions in the presence of the enemy, which are generally decisive, if done with precision and exactness, which you dare not even attempt, unless you are certain of being able to execute them. Generals ignorant in this sublime and delicate part of war, are incapable of changing their plan
according

according as new circumstances arise, which always do arise, because as the enemy approach, they very justly fear to make any movement in his presence, as they do not know whether they have ground, or time enough, to execute this or that manœuvre, though convinced of their necessity."—See *Lloyd's Preface to the War in Germany.*—See also *James's Military Dictionary, second edition,* under the head *coup d'œil:* the article would be given here but that it is presumed that the book itself is in the possession of every officer. To conclude this interesting subject in the words of M. Guibert, " commanding officers cannot apply too much to the instructing those under them in the *coup d'œil,* to the improving of their own, or in preparing it for those illusions, which the variety of objects often produces; consequently manœuvring their troops sometimes on even ground, at other times in rough and inclosed countries, where many impediments exercise his genius, and inure the soldier to the toil of war."—*Gen. Ess. on Tac.* vol. i. p. 295.

NOTE

NOTE (6).

" Arms are called *arma* in Latin, by the Germans *wappen*, or *clenodia*, and by the French *armoiriès*. Arms, with us, means all sorts of warlike instruments offensive and defensive. From wappen, it is presumed, comes our word weapon, and wapentakes, the latter being places of rendezvous where men assembled in arms."— See *Guillam's Display of Heraldry*, c. i. p. 10. " Arms is a French expression to signify the different qualities of corps, such as horse, foot, dragoons, artillery, irregulars, &c."—*Cadet*, p. 20. In this latter sense the word arm is used in the text. M. Guibert applies it in the same manner: thus he says " artillery is the third arm of an army."—Vol. i. p. 313. " Arm, from the French *arme*, signifies a distinct branch of the army. Cavalry is an *arm*. Artillery is an *arm*, &c.—*Gen. Hist. on Tact.* vol. ii. p. 145.

NOTE

NOTE (7).

On this important subject, not perhaps sufficiently attended to, M. Guibert has the following observations :— " When troops (infantry and cavalry) are united by a reciprocal protection, to draw the greatest advantage from their assistance, it is necessary that artillery officers should know the tactic of troops, according to its internal detail : the least they should be acquainted with, should be the result of the principal movements, the change they convey to the order of troops, and the mischief or succour they receive, on more or less favourable occasions, from the execution of artillery when well posted accordingly.

" There are stronger reasons why, in like manner, the commanding officer of infantry and cavalry should know how to command artillery : if ignorant of the interior details of its construction, he should be acquainted with the result of them ; the range of the different pieces, when pointed or posted in such or such ground, &c. the injury which one direction occasions to troops, and the service any other conveys. Deficient

ficient in this knowledge, either he knows not
how to employ artillery with propriety in his general dispositions, or he will be necessitated to
apply, in a confused manner, for the manœuvres of artillery, to an officer of the corps, who,
likewise (perhaps not having extended his knowledge beyond the mechanical use of his piece),
will not be able to dispose it, so as to complete
the views of the general ; or, in short, he would
perhaps act contrarily, through ignorance, to the
good dispositions.this artillery officer had made."
Gen. Essay on Tactics, vol. i. p. 337.

NOTE (8).

On the subject of exercising the sister arms
(infantry and cavalry) together, the enlightened
author quoted in the last note, remarks :—
" Since so great an affinity exists between these
sister arms, their reciprocal jealousy should be
destroyed: let them regard each other as inti-
mately connected, and of equal use in war.
'Tis true, that the infantry can act and engage
without cavalry : but its progress would be slow
and

and dangerous; it would be harrassed without
end, continually exposed to the loss of subsist-
ence, and few of its operations would be carried
on with alacrity. The cavalry, without infantry, would perform nothing decisive; it would
find no spot to station itself in, the most trifling
obstacle would retard its motions: at night it
would be in danger of its safety."—*Gen. Essay
on Tactics*, vol. i. p. 260.

" Exercising of them (cavalry and infantry)
together," says the author of the *Cadet*," would
divest them of those prejudices which they too
commonly entertain against each other, as they
are for ever (during peace) ignorant of each
others connection: whereas they should know
how essential, and how necessary their mutual
assistance is to each other."—*Cadet*, p. 72.

NOTE

N O T E (1).

THEREFORE it is laid down in the *British Tactic for the Infantry*, as a general rule---" That a well-drilled soldier be placed at the head of a column, horse or foot, and to march at the regulated cadenced step, the rest following at loose files, and at their ease, the pivots only strictly preserving their distances in all situations."---*Rules and Regulations*, p. 369-70, edit. 1803.

And so well aware is His Royal Highness the Commander in Chief, of the necessity of habituating the troops to the cadenced step, that the orders on this head are express and peremptory :---" At the head of every column, whether composed of infantry or cavalry, a well-instructed non-commissioned officer must march. He must keep the regular step of the slow march: upon this man the regular pace of the column will depend."

It

It is further mentioned, " That two non-commissioned officers should be appointed for this purpose, who must relieve each other."— *Gen. Reg.* p. 41, 42.

Perhaps it might be advisable to have ten men in every regiment trained to this particular and most essential purpose, so that their step in all sorts of ground, and in every situation, might be depended on : they should also possess a correct *coup d'œil,* to enable them quickly to· take points, place camp colours readily in a line, or at a right angle, &c.; these might act as camp colour men, and be denominated guides or regulators of march, and have some privilege above the other men, to excite them to steadiness and sobriety of conduct. In fact, nothing but an unwearied attention to habituate the soldier to the regulated step can ever insure accuracy of movement; 'tis especially necessary to the column of route : how essential it is esteemed, in the highest disciplined armies on the Continent, may be seen by consulting the *Tactique Prussienne,* and the *Regulations for the French Infantry,* under the head *L'Ecole de Bataillon.* Its advantages were well known to the ancients.

" The

" The Greeks and Romans," says M. Guibert, " had a regular cadenced march; but
what its measure was, we are ignorant of. It is
in our time that the use of the cadenced measure of the march was established in Europe, I
may say discovered, as for many ages it slept in
security. Marshal Saxe esteemed it as the most
interesting circumstance, and which ought one
day to make a great epoch in the improvement
of the tacties. This great man seemed as
though he had the skill in foretelling the revo-
lutions which were to be made in the principles
of this science, when he wrote (in his *Reveries*),
that all the mystery of the tactics were to be
found in the legs."—*Gen. Ess. on Tactics*, vol. i.
p. 136.

Speaking of forming troops for action, Marshal Saxe says :—" I shall begin with the march,
which subjects me to the necessity of first advancing what will appear very extravagant to
the ignorant : it is, that notwithstanding almost
every military man frequently makes use of the
word *tactick*, and takes it for granted, that it
means the art of drawing up an army in order
of battle, yet not one can properly say, what
the ancients understood by it. It is universally
a custom amongst troops to beat *a march*, without

out knowing the original or true use of it; and it is universally believed, that the sound is intended for nothing more than a warlike ornament.

After enumerating the many evils attending the want of regularity in marching, he goes on to say (p. 16), " that the way to obviate these inconveniencies is very simple, because it is dictated by nature ; it is nothing more than to march in *cadence*, in which alone consists the whole mystery; and which answers to the military pace of the Romans: it was to preserve this, that martial sounds were first invented, and drums introduced : and in this sense only is to be understood the word *tactick*, although hitherto misapplied and unattended to."

It was the cadenced step that enabled the Romans to maintain their military pace, with which they marched twenty-four miles, equal to eight of our leagues, in five hours. " Let us try," says he, " the experiment upon a body of our infantry, and see whether they will be able to perform as much in the same space of time." He looked upon it as a matter of the highest importance, to be enabled, by the assistance of the cadence, to be able to encrease or diminish

K   the

the rapidity of a march, during an engagement.
And he adds :—" It will be no difficulty to
prove, that it is impossible to keep the ranks
close, or to make a vigorous charge without it."
Reveries, p. 18. ‹

" So essential," says the Marshal, " is marching, that the army which marches best must, if
the rest is equal, in the end prevail." To the superior marching of Turenne's armies he imputes
all their victories; the infantry of France, under
Turenne, was reckoned to be at that time the
best in Europe, and in the best condition for
service; " had it been otherwise," says Saxe,
" how was it possible for him to make such
long and glorious campaigns? There is one
march in particular, that we all know was made
by M. de Turenne, which at present would be
impracticable for us to perform in so short a
time."—*Memoirs sur l'Infanterie*, p. 11.

The author of the *Cadet* says : " I believe
M. Saxe means the march from Hochfeld to
Wisloch, about thirty leagues in four days."—
Cadet, p. 7.

" To besiege St. Venant, a town upon the
Lys, in the county of Artois, the Viscount de
Turenne

Turenne marched twenty five leagues in three days." See *Life of Turenne*, vol. i. p. 298.

It is also to be considered, that the *foot* soldier in Turenne's time, was much heavier armed than he is at present; the musket and accoutrements were more weighty and cumbersome than they are now : and the pikemen, who formed a third of each battalion, wore defensive armour, the demi-cuirass, and the casque, armet, bourguignotte, morion, or pôt, for they all signified nearly the same thing: the latter, with the corset worn by the pikemen, were both proof against the pike, and the cut of the sabre. The casque or pôt alone weighed from 18 to 20lb.

" Le bourguignotte, que se nomme aussi, armet ou morion, est un pôt qui accompagne ordinairement les corselets des piquiers ; ces corselets & ces pôts sont à l' epreuve de la pique & du coup d'épée." See *Memoires d'Artillerie, par M. St. Remy*, tom. i. titre xvii. p. 297.

" A Roman soldier's arms weighed," according to Marshal Saxe, " above 60lb. and it was death to throw away any part of them in action."—*Reveries*, p. 31.

A foot soldier in the British Service does not

K 2　　　　　　carry,

carry, including his great coat and necessaries complete, so much as 50lb. weight, in marching order, according to the following calculation :---

	lb.	*qrs.*
Great coat and straps - - - - - -	5	2
Canvas painted knapsack with full necessaries, according to regulation - -	12	0
Firelock and bayonet, with sling, - -	10	3
Pouch and belts, black - - - - -	4	0
Sixty rounds ball cartridge - - - -	5	3
Total	38	0

The arms were what are called East-India arms. The Tower arms weigh 14lb.

NOTE (2).

Troops accustomed to parade and manœuvre on level ground, might find themselves embarrassed, if in broken or uneven ground they should at any time be obliged to attempt a march in line, or a deployment from column into line, so as to be obliged to give battle in such situations. In Jersey, over the broken sand hills at Grouvill, a regiment has been seen

to

to march 6 or 700 yards in line, without any deviations, and under such circumstances, and in marching order, ankle deep in hot sands, have gone through the Eighteen Manœuvres with as much precision as if on a plain and firm ground. This could only be the consequence of men being well grounded in the true principles of marching, and under the direction of excellent officers. Our masters in the art of war understood this well; there was no situation of ground a Roman army could be in, that it was not prepared for battle, so admirable was their discipline. The following quotations may serve to illustrate it, and to recommend it to our practice.

" As engagements may happen in all kinds of places, the Romans not only exercised their soldiers in the open plain, but in defilés, narrow straights, and passes, in which it would be difficult to march, and preserve their order; that being obliged to fight in such places, they might by this habitude be less exposed to the confusion, which the novelty might occasion."--*Mar. Puysegur Traité de la Guerre*, p. 60. *Cadet,* p. 54.

" It was not in the plain alone, that the Romans paraded, they ordered their soldiers into

difficult

difficult and uneven places, where they should
ascend or descend a mountain, to prepare them
for all kinds of accidents, and to form them for
all the different movements that the *site* of the
ground might require." *Veget.*—*Ins. Militaires.*
—*Culet,* p. 55.

When the great Turenne, in 1658, was laying
siege to Dunkerque, the Spaniards came to its
relief; they encamped on the downs or sand-
hills near it. The Viscount marched out to
meet them the day after they appeared, his army,
" without reckoning those which guarded the
baggage and the trenches, amounted to 6000
horse, and 9000 foot. The Spanish army under
Don John and the great Condé consisted of
6000 foot and 8000 horse, without cannon.
Turenne's infantry was in two lines: the first
consisted of 10 battalions, and 28 squadrons, 14
in the right wing and 14 in the left, with the
cannon at their head. The second line consisted of 6 battalions and 20 squadrons, 10 to the
right and 10 to the left. Four squadrons of
gendarmes (heavy armed horse) supported the
infantry, and 6 squadrons of reserve were posted
at a good distance behind the army, that they
might be ready to succour the besiegers in case
of a sally during the battle. The first line was
about

about a league long, reaching from the sea to
the canal of Furnes. The two lines in which
the battalions and squadrons were drawn up,
seemed to be quite straight, notwithstanding the
unevenness of the ground." The two armies
were not above a quarter of a league distant
from each other. " The French, to avoid *breaking their* ranks in that uneven ground, advanced
so slow, that they were *three hours* in going the
quarter of a league that was between the two
armies," so essential did this great man esteem
the march in line, and in correct dress, so as to
attain the enemy at once on all his points. This
was the famous battle of the Downs fought on
the 14th June, 1658, in which the Spaniards
were defeated by the French and English; Dunkerque surrendered ten days after. ' The Viscount, after this glorious day,' says the historian, ' wrote the following short letter to his
lady: "The enemy came to us, and God be
praised! they have been defeated; I was pretty
busy all day, which has fatigued me; I wish you
a good night; I am going to bed."—*Life of
Turenne by the Chevalier de Ramsey*, vol. i. p.
317, a work of distinguished merit, highly in-
structive to every military man, and particularly
worthy the attention of the young officer, and
of all who wish to make the army their profes-

K 4   sion.

sion. It is to be recollected that in Turenne's time the foot soldier occupied three feet in front, and five in depth.—*Montecuculi,* l. i. c. iv. p. 97. But when near an enemy they closed to 18 inches.—*Morden's Art of War.*

N O T E (3).

IT has been well remarked by the very able author of the Rules and Regulations, " that the necessity and importance of division marching, is not sufficiently impressed on the minds, or sufficiently attended to by British officers; yet it is only in that order, that a battalion should at any time perform its marches, that the columns of an army should be permitted to move, that an enemy should be approached; and that safety can be insured to the troops in their transitions from one point to another."—*Rules and Regulations,* part iv. p. 368.

Yet the old mode of doubling the files from line, then facing and marching on that front, is still practised by some regiments. Thus suppose the regiment stands three deep, by doubling the

the files, and facing (to the right suppose) it
marches on a front of six files; if in this case its
head was threatened, so that it was necessary
to form line to that front, it could not be done,
as the files are mixed and confounded together;
to remedy this defect, the battalion must halt,
front, and the files be reformed, to place the battalion in its natural form. But if a column of
sections is in march, and it is necessary to
form line on its leading section, the operation is
perfectly simple and easy, the sections move up
into divisions, platoons, or grand divisions, and
finally into line, without any risk of confusion;
so if the column of sections was to form line to
its proper front, it is done instantly and accurately by the wheel up of its divisions, an advantage that the column of mixed files cannot have,
unless that be supposed, which cannot be granted, that in its march, each soldier (in the front
rank particularly) should keep so exact a distance from the man before him, that at the
words *halt, front,* there would be room for the
file to come up, and no more; for if there was
more room than was exactly necessary to admit
the man, the line would have its files too loose,
and must close to its center to correct itself; or
if there was not exactly room for the files that
came into the front, the line would be too much
straitened,

straitened, and must extend before the soldier
would have the proper distance in which he
could handle his arms.

Much more might be said on this subject,
but a moment's reflection may shew any man,
though ever so little acquainted with military
affairs, that a column formed by doubling of
files, is radically bad; and on a throrough con-
viction of its being so, has accordingly not been
thought deserving of a place in the *Rules and
Regulations for the British Infantry*; why it
is permitted by officers who should know better,
is as incomprehensible, as that they should occupy the time of the soldier in teaching him
that which can never be of any use to the service. And it may perhaps be safely affirmed,
that wherever the strict letter of the *Rules and
Regulations* has been in any instance departed
from, that such deviation has always been for
the worse. The method of doubling the files,
&c. may be seen in *Morden, Bland,* and in
Simes, p. 186.

NOTE

NOTE (4).

Captain Tielk mentions this, in his *Field Engineer*, vol. i. p. 209 :—" And an admirable use might be made of a plan of this sort, in an army like ours, that changes its quarters at least once in every year: what a fair opportunity of obliging the column, or any of its divisions, to be led at the regulated pace, leaving nothing to the caprice of the conducting officer. Thus, suppose a regiment, or any division of it, was to march, the commander would place one of his drilled men at the head of the column, and noting the hour of departure to a minute, give the word *March;* changing the leading man as occasion required (which would perhaps be every half hour, as no man can be well supposed to march for a longer time with a stiff knee), and proceed to where he first halts; then mark the moment of arrival, the number of miles marched, and the moment of departure, and so on, till arrived at the end of the day's march; the whole number of miles should then be noted, and the time spent in the several halts deducted, would give the exact rate of march. Then should be noted also, what state his men were in, when ar-

rived

rived at the end of the day's march; whether fatigued, or otherwise; how many fell behind, and on what account; whether their feet were blistered, or from what other cause they were disabled from marching. It should be mentioned also, at what time, to a second, the baggage set off; of what number and description of waggons, &c. it consisted; what halts they made, and the exact moment of its arrival at the end of the day's march: this journal should be continued till the troops had arrived at their final destination; then would be seen, at the end of each day's march, how far the order was complied with, for the column to be conducted at the rate of three miles in 75 minutes (*General Regulations and Orders,* p. 62); and this journal should be signed by whoever commanded, and was actually with the division: this journal would require nothing more than attention and correctness.

The nature of the country passed through should be carefully noted also; whether open or enclosed, hill or dale, woodland, lawn or meadow, pasture, arable or waste, heath, moor, or bog; the roads, whether good or bad, sandy or stoney, gravelly or clay; what towns, villages, or hamlets were passed through, and their

situations

situations as military posts, and what were in
sight, how far from the main road, and on what
side, and their situations as far as might be
guessed; as also what bridges, fords, mills,
church yards, strong houses, stone walls, &c.&c.
what rivers, rivulets, or brooks; the weather,
whether hot, cold, or temperate, frost, snow, or
wet, &c. Any officer, who could draw (and
surely such should be encouraged), might sketch
whatever was remarkable, in a military point of
view, according to the idea laid down in the
ingenious treatise on *Military Plan-Drawing*,
by Captain Malorti de Martemont.

All observations made, might be arranged
under different heads, on printed forms, and
transmitted to the proper department; its be-
nefits would be incalculable : what a treasure of
useful information would be obtained, by the
War Department, at no expence! what an opportunity for officers to exercise their *coup d'œil*,
and arrive at perfection in it ! and by obtaining
an exact knowledge of one country, be led to
form a correct idea of any other in which they
might be employed.

The French have availed themselves of talent
of their officers in this particular. Their War-
Office

Office contains plans and drawings of every
country in Europe; the interior of every pro-
vince is laid open to their view.

This interesting subject might be pursued
much farther; all that is presumed here, is a
hope, that some one better qualified may yet un-
dertake to place all its advantages in a clearer
point of view; as a plan of this sort would be
conducive to the benefit of the service in general, materially tend to facilitate the movements
of troops, as the hour of their departure from,
and arrival at, named places, in given spaces of
time, might always be depended on : it would
habituate them to the regular cadenced step,
and would add greatly to the ease and health of
the soldier, who would thus arrive at his quarters without hurry, confusion, or fatigue, which
sometimes happens from the unequal paces at
which it pleases their commander to lead them.
And it is laid down in our tactic, as a rule not to
be departed from, " That the column of route
should never be suffered in any case to be conducted except at the cadenced step, as nothing
else can ensure the troops from being hurried,
especially when cavalry leads."—*Rules and Regulations*, p. 369.

What

What has been said here, applies when either cavalry or artillery are on their march : the same returns, &c. as have been proposed for the infantry, should also be made by them. See also, on this subject, *Plan of a Military Common-place Book*; and James's *Military Dictionary*, under the head SKETCH, edit. 1805.

NOTE

NOTE (1).

(Remark E.)

CAPTAIN TIELK observes, "That the object fired at, should always be within point-blank shot, as all those fired above the horizontal range are uncertain; we have not been able (says he) to ascertain the true distance of point-blank shot, nor that fired with elevation."—*T. G. Tielk, Captain of Artillery in the Service of H. S. H. the Elector of Saxony.*

NOTE (2).

" For great bodies, consistent with perfect order, cannot move with the same degree of rapidity as smaller ones."

NOTE

N O T E (3).

For it is a rule laid down in our cavalry tactic, " That a squadron should never be so much hurried as to bring up the horses blown to the *charge*, and therefore an allowance proportionable to their extent, and the length of ground they have to go over, must be made in conducting them."—*Cav. Reg.* §. 11. p. 31.

N O T E (4).

Ricochet, or duck and drake fire, is the most destructive that artillery can make use of with round shot; " for the balls will not only take place, but likewise create fear and disorder among the enemy's troops; and if point-blank shot strike a platoon, or squadron, and take off a few men, they neither create terror nor confusion, being only observed by their effect: but those which the soldier sees at 100 paces in front, bounding towards him, will occasion a wavering and disorder in the ranks, by his en-

L   deavouring

deavouring to avoid them." See *Tielk's Field Engineer;* also *Le Blond's Elements of War,* under the head of *Ricochet, or rolling and bounding Shot,* p. 23. Ricochet shot, according to Le Blond, was invented by M. de Vauban, and used by him, in 1697, at the siege of Aeth.

NOTE (5).

(Remark E.)

With this view, when the enemy's line advances, the artillery officer would not fire directly at them; " but he would take aim rather before them, and in the retreat beyond them: for in the first instance, the soldier believing that he approaches danger and death as he advances, will be discouraged, and slacken his pace; and in retiring, the shot will be thrown beyond them as well as among their ranks, this will augment their terror and confusion, they will retreat slower and in greater disorder, and more prisoners will be taken."—*Tielk's Field Engineer.*

NOTE

NOTE (1.)

(Remark F.)

See *Rules and Regulations*, p. 375, edit. 1803.

Captain Tielk observes, that " if a corps of infantry, formed three deep, advances direct on a gun, a shot can only destroy three men; but it is otherwise, if taken in column, for having a greater depth of troops to fire on, more execution would be done. At the battle of Zornsdorff (1758), one cannon shot mowed down forty-two men belonging to the 2d regiment of Russian grenadiers. Cannon should not be discharged along the line, but collected in strong batteries, not fired upon an enemy's front, but obliquely, and *en écharpe*."—*Tielk's Field Engineer*.

NOTE

NOTE (2).

(Remark F.)

The excellent author mentioned in the last note observes, " That troops can march, without running, from 100 to 120 paces in a minute; therefore when a battalion has marched in line from 200 to 250 paces, exposed to a cannonade, it will have received twelve, or at most fifteen shots from each piece, only the lesser number of which will take place, for the line continually approaching, the bombardier is obliged to take a different aim at each fire."—*Tielk's Field Engineer.*

See *Rules and Regulations,* part iv. p. 375.

———

N O T E (1).

OUR light dragoons seem to be improperly so denominated; they should, perhaps, rather be styled hussars, as they approach nearer to that description of horsemen than to dragoons.— Hussars are of Hungarian origin : their cavalry are called hussars, as their infantry are hey-duques. " Les heyduques," says M. Monte-cuculi, " sont l'infanterie des Hongrois, & les Houssarts leur cavalerie."—*Mem. de Montecu-culi*, L. iii. C. iii. p. 423.

According to Mr. Guibert, some French officers, arriving from the wars in Hungary, who had seen the utility of the irregular troops of the Turks and Hungarians, brought a few hussars with them : this gave the idea to Marshal Lux-embourgh, in 1692, of raising the first regiment of French hussars, called *Mortagni*; the second was raised by Marshal Villars. The Elector of Bavaria made a present of a third to the King." See *General Essay on Tactics*, vol. i. p. 301.

Another

Another French military writer says, that the
first hussars in France were some deserters from
the army of the Emperor, and not being at first
trusted to act in the French armies, they were
retained by some of the great officers, as appen-
dages to their state (a custom not yet quite exploded, as we sometimes see one dressed and
whiskered, as a hussar, standing behind a coach);
however, the author observes, as these deserters
increased, they did not like to be thus used, but
wished to be employed in the army; and one
bolder than the rest, ventured to speak to Marshal Luxembourgh, begging him to put their fi-
delity and valour to the proof: he got them
formed into a company, afterwards they increased. At first they had no discipline, and
were only employed in attacking convoys, &c.&c.
They rode in confused masses, but darted like
lightning on the prey that each had in view.
When they went out, they seldom returned
empty handed; " puis qu'ils ne revenoient ja-
mais les mains vuides." Their pay was regulated
by an ordonnance, 20th April, 1722.—See for the
above, *Ecole de Mars*, L. v. p. 3.

They answered nearly to the description of
the irregulars given by Montecuculi. The
cossaques of Poland, or Zaporuwisch, those
who

who inhabit the Upper Volhinia, the Ukraine, Kiovia, or those under the Turkish government, inhabiting the banks of the Nieper, or in the country of the Tartars of Oczakow; these last are of the Greek religion, good soldiers, and great robbers. " Ils sont de la religion Greque, bons soldats, & grands voleurs."—*Mem. de Montecuculi*, l. ii. ch. iii. p. 331.

Lewis the Fourteenth had considerable numbers of hussars and dragoons. In 1740, the next heir to the possession of Charles the Sixth, the beautiful Maria Theresa, was obliged to fly for succour to the arms of her Hungarian subjects: " Then," says M. Guibert, " appeared in Germany the military of that nation, Transylvanians, Croats, and other irregular and undisciplined troops of the House of Austria."

Marshal Saxe appointed Uhlans, and regiments were then raised that were called light troops: at the same time the King of Prussia made great levies of hussars and dragoons, to act against the Arrier bands of Hungary. In 1756, this reciprocal increase of light troops was carried to a much greater extent." Guibert, vol. i. p. 301 and 302.

The

· The reader may see in *Lloyd's Rhapsody*, c. 6, p. 65, how prodigiously light troops increased in Europe.

The first regiment of light horse that was raised in England was in 1746 : they used the straight sword, and not the sabre. — *Grose's History of the British Army.*

" There were a sort of light cavalry in France very early, that is, they were not so heavy as the gendarmerie. The first that appeared there, under any rule or form, were those which the Bannerets, and other principal nobles, led with them to the wars, under the name of archers. But as these cavaliers were all vassals, and even in the pay of those who furnished them, and that they were so dependent on them, that they would acknowledge no orders but theirs, — Charles the Seventh had them formed into companies, subjected to his regulations. Louis the Twelfth did the same, forming them into independent companies. Francis the First found them useful, and augmented them : and under Henry the Second, the nobility being much weakened, these troops were officered and regimented; so that the cap-a-pie armour having been abolished, under the reign of Henry the Fourth,

Fourth, there was no difference between the light cavalry and the gendarmerie."—*Ecole de Mars*, l. v. p. 3.

Marshal Brissac, in his wars in Piedmont in 1600, raised the first regiment of dragoons. The Spaniards were the first people who copied the French; in a short time afterwards the remaining powers in Europe raised dragoons.— The dragoons of M. Brissac were more properly infantry mounted, having continued the use of the musket and pike; they fought on foot or on horseback, but generally on foot; they were armed with swords, musquetoons, and half pikes: they were sometimes placed in the open spaces between cavalry regiments, to fire on the infantry; but they were chiefly useful in seizing expeditiously on a post, forwarding the march of their own army, or impeding that of the enemy in an inclosed country, or in narrow passes; for which purposes they had mattocks and shovels, and acted as pioneers; they had the worst horses selected from the cavalry, so that if they were obliged to quit them their loss might be less; they wore neither boots or spurs; when they fought on foot, they tied their horses together two by two: sometimes they were classed in divisions of eleven each, while ten fought

on

on foot, the eleventh held the horses. See for the above, *M. Guibert's Essay on Tactics,* vol. i. p. 301—*Grose's Hist. of the Brit. Army,* vol. i. p. 110—*Mem. de Montecuculi,* l. i. c. ii. p. 16 and 17.

The only mark that our light dragoons yet retain, that distinguishes them as such, is the carbine; of which it would be better perhaps to disencumber them intirely. It is not easy to suppose a situation in which the carbine (such as they now carry) could be of much, indeed of any use: to oppose its fire to that of musketry would be absurd; or to use it against cavalry would be equally so. " All fire-arms (says M. Guibert, vol. i. p. 262) made use of, are ill adapted to the regulation of cavalry: but if fire-arms are admitted for the use of cavalry, they are not supposed to be of use when the men are mounted;" for in no case would he admit of firing on horseback. Dragoons, however, properly so called, he admits, may have fire-arms, for this purpose, that " when there is a want of infantry, and the horsemen being obliged to dismount, to occupy the head of a defile, or some other useful purpose, it is necessary that they (dragoons) should be armed with a car-a carabine and single pistol."—*General Essay on Tactics,* vol. i. p. 262.

Our cavalry Regulations say, "That the light horseman skirmishing, must not fire on the halt, but on the move;" at the same time he is cautioned (very properly) to take the greatest care not to hit, or burn the horse's head, or at that time to spur him."—*Cav. Reg.* p. 353.

Indeed so little effective did the great Saxe suppose the fire of cavalry to be, and so little was it their forte, that he deemed it superfluous to say, that they should not fire, as it is a matter that every one knew so well. His words are, "Il est, tout aussi superflue de dire qu'elle ne doit pas tirer, & que le feu n'est pas de son ressort, tout le monde le sait assez."—*Esprit des Loix de la Tactique, ou Notes de M. le Marechal de Saxe, Commentees par M. de Bonneville, tome premier,* p. 145.

If what has been said against the fire of cavalry be just, it would appear, that this arm does not properly belong to the cavalry; and the horseman should depend for success alone on his valour, his sword, and the celerity of his movement.

NOTE

N O T E (2).

(Remark G.)

The observation of Captain Tielk, in his *Field Engineer*, on this head, deserves attention.— " The number of men killed," says he, " will not gain the victory, nor even abate the courage of the combatants : the soldier knows that some of his comrades must fall, and he has neither inclination nor leisure to count them: to deprive him of his courage, and to turn his attention to his own safety, he must be struck by some extraordinary incident; let him see whole ranks and platoons swept off; he will be seized with horror and dismay at the scene of death around him; he will be deaf to the voice and command of his officer, and no human power can prevent his immediate flight. For which reason," says he, " artillery should not be scattered along the front of the line to fire direct at the men, but placed in batteries for the enfilade; therefore great care should be taken to avoid that destructive fire, the effect of which on a column is so dreadful."

NOTE

(157)

NOTE (3).

(Remark G.)

" The cavalry," says M. Guibert, " when charging, has no doubt a force in its shock ; but this shock is the only production of that share of velocity by which it is put in motion, and the sum of that mass which is centered in the first rank. That portion of mass which is centered in the following ranks adds no force to the first.—*Gen. Essay on Tactics*, vol. i. p. 268.

This appears to be correct, as horses cannot push one another forward, as men might be supposed to do: and in the full gallop, the ranks must be four feet asunder. " Half a horse's length between the ranks must in the charge be most attentively adhered to, it gives but bare room for the horses to strike their legs in speed; from inattention to this precaution, the front rank horses are miserably cut and trampled on." *Light Horse Drill, by a Private of the London and Westminster Volunteers*, p. 36, 3d edit.

Were the infantry well informed on these points (as no doubt they should be), a column of cavalry, however deep it might appear approaching

proaching to the charge, would have no more
terrors for them, than if they came on one or
two ranks; for if the fire of the infantry repulsed the first rank, they would have soon an
intrenchment of men and horses before them,
" so that was 100 squadrous to advance in a
successive manner on the same point, they would
only be disordered by stumbling on the slain,
and perhaps expose themselves to the same fate
as the prior ones."—*Gen. Essay on Tactics*, vol. i.
p. 290.

But on this principle, it might appear, as if
cavalry can derive no advantage from its depth;
yet Marshal Saxe says, " It is an error to believe
that the files of cavalry do not draw from their
depth the advantages that the infantry do. I
say the first ranks feeling themselves pressed, and
followed by a train of many others, in acquiring
more vigour and force, it cannot be otherwise
than that they are pushed forward, for they cannot go back." His words are, " C'est une erreur de croire que les fils de cavalerie ne tirent
point de leur hauteur l'avantage qui en résulte à
l'infanterie ; & je dis que les premiers rangs qui
se sentant pressés & suivis par une suite de plusieurs autres, en acquierent bien plus de vi-
gueur & de force : il faut, absolument qu'ils
poussent

poussént en avant, car ils ne peuvent pas reculer."—*Esprit des Loix de la Tactique ou Notes de M. le M. de Saxe, &c.* (already quoted) *tome premiere*, p. 144.

NOTE (4).

(Remark G.)

The method of firing by whole battalions to repel the charge of cavalry, as mentioned, may seem to be contrary to the tactic for all the infantry of Europe. " The chief object of fire against cavalry is to keep them at a distance, and to deter them from the attack; as their movements are rapid, a reserve is always kept up."—*Rules and Regulations,* part iv. p. 349: therefore the firing by platoons is what the soldier is practised in, to keep the cavalry at a distance, and to afford also what is called mutual protection. But could any thing so effectually deter cavalry from advancing, as to receive one sharp fire, and seeing the infantry coolly preparing to give them a second, when every shot could be fired with destructive certainty? Platoon firing is productive of noise, smoke, and confusion:

confusion : but if the following obfervations of
an officer of great talent be just, it would appear, that a battalion could not make more discharges in a given time, from firing by its
fractional parts, than by vollies given by the
command of a steady officer. " The course of
this fire (by platoons) begins with the fourth
platoon, and ends with the grenadiers; the mutual protection exists during half its course,
after which it decreases, and is at length reduced to the fire of a single platoon; and such
is the condition of the battalion, while this fire
is continued. If the enemy, whom we shall
suppose only fifty or sixty paces distant, should
in those circumstances advance to the attack
sword in hand, it is certain they would receive
only the fire of three or four platoons, and find
the rest defenceless and busy loading ; mutual
support and protection therefore is not to be
found in the manner of employing the fire in
question ; and the general fire, notwithstanding
it defects, is infinitely better : for beginning by
this, the whole corps would be as soon ready
for the second discharge as the fourth platoon,
and the two discharges would be equal to the
fire of eighteen platoons: moreover, the second
discharge of the fourth platoon makes but the
fire of ten ; therefore the general fire is to the
platoon

platoon fire as eighteen to ten, which is almost two to one."—*Elements of Military Tactics, by the Sieur B—*

It must be observed, that a battalion of French infantry, which is what has been here spoken of, consists of eight battalion companies, and one of grenadiers; in all, nine companies."—*Réglement concernant l'Exercise et les Manœuvres de l'Infanterie,* p. 3.

Marshal Saxe was no advocate for firing in general, unless in certain situations, where, as he says, it may be necessary, such as in inclosures and rough grounds, and against cavalry: " but the method," says he, " of performing it, ought to be simple and unconstrained. The present practice is of little or no effect; for the men are so taken up by the attention which they are obliged to pay to the word of command, that it is impossible for them to fire with any certainty: how is it to be expected, that after they have presented, they can in such a position, retain an object in their eye, till they receive the word to fire? The most minute accident serves to discompose them, and having once lost the critical instant, their fire afterwards is in a great measure thrown away: the

M   strictest

strictest nicety and exactness is required in levelling, insomuch, that any movement of the firelock, when presented, although even imperceptible, is sufficient to throw the ball considerably out of its true direction; to add to which, their being kept in a constrained attitude, will naturally make them unsteady: these and other inconveniences totally prevent that execution which might be expected from small arms."— *Reveries*, p. 33.

The Marshal had not a very high opinion even of the efficacy of a general discharge, though made at so small a distance as thirty paces, of which he gives an instance, at the battle of Belgrade, where two intire battalions of French infantry were attacked by a body of Turkish horse: " I saw," says he, " the two battalions present, and give a general fire upon a large body of Turks, at the distance of about thirty paces; instantaneously after which, the Turks rushed forward through the smoke, without allowing them a moment's time to fly, and with their sabres cut the whole to pieces on the spot. The only persons who escaped, were M. de Neuperg, their commander, who happened luckily to be on horseback; an ensign, with his colours, who clung to my horse's mane, and incumbered me

not

not a little; besides two or three private men. I had curiosity enough to count the number of Turks, which might be destroyed by the general discharge of the two battalions, and found it only amounted to thirty-two; a circumstance which has by no means increased my regard for the firings."—*Reveries*, p. 21 and 22. And in a note (p. 22), on the same subject, it is observed : " The quickness with which the Prussians load, is an advantage in one respect, as it engages a soldier's attention, and allows him no time for reflection in marching up against his enemy: nevertheless, it is an error to imagine, that the five victories, which they obtained in the last war, ought to be ascribed altogether to their firing, because it was remarked, that in most of those actions, they lost more men from the fire of the enemy, than the enemy did from theirs."

The effect of a general discharge, made by well disciplined troops, and by the word given by a cool and steady officer, may be conceived by the following fact :—At the battle before Alexandria, so glorious to the British arms, the 28th Regiment, commanded by Lieut. Col. Chambers, was posted in a *fléche*. On the dawn of that memorable day, they were attacked by

a body

a body of Copts or Greek grenadiers, who resolutely persisted in storming the front of the
work; many had entered, and were killed:—
while they were thus furiously engaged, a party
of French cavalry (about thirty in number) made
a determined attack on the gorge of the work :
in this situation, vigorously attacked in front
and in rear, the commander ordered a part of his
men to face about; when they gave, by his
word of command, so well-directed a fire, that
every man and horse of the intire of the party
were knocked down, and either killed or disabled by that one discharge.

NOTE (5).

(Remark G.)

The following observations on the oblique fire,
it is presumed, will not be thought wholly unin-
teresting : " The effect of the oblique firing,"
says the ingenious author quoted in the last
note, " is not sufficiently known. The Chevalier Folard declares it to be the most terrible of
all fires, but without explaining himself.—
Troops

Troops cannot fire but in a direct line before them. An oblique fire can mean no other than where the direction is oblique, without regard to the enemy's front." But the reason why this fire should be more destructive any other is quite simple: if we place ourselves facing the center of the battalion, and throw our eyes directly on it, we shall see a number of vacant spaces, and consequently openings for the passage of many shot; if afterwards we direct our eyes from the same point to the wings, we cannot perceive the least opening through the files, and every ball fired in that direction must take effect. This fire therefore is more destructive than any other, as each shot in that case must necessarily wound somebody."—*Elements Mil. Tac. by the Sieur B——, p. 79.*

M. Guibert observes, that the oblique fire is the best to be used, even if the front of the enemy was of the same extent only as your own. " For (says he) on account of the fires being convergent, and assembled to one point, they would become more destructive."—Vol. i. p. 169.

" The Prussian and Austrian infantry have a method of changing the firelock to the right and left shoulder, by which means they procure,

on

on all occasions, a flank and oblique fire."—See the work above quoted, vol. i. p. 169. And it is added, in a note : " This kind of fire was practised last war (1756) with great success, particularly at Sweidnitz and Torgaw; since which time, the Austrians (who are the faithful imitators of the Prussians) have introduced it into their army."

NOTE (6).

(Remark G.)

It must be confessed, that the situation of a soldier in the front rank, waiting to receive the attack of a horseman, is as helpless and as pitiable as can easily be conceived; he does not bend his knee to the ground (a situation from which he never expects to rise) until he has (probably) fired his last shot: all he had expected from his musket as a missive weapon is at an end; and in this situation he must endeavour to use it as a sort of barrier against the horse, and sloped so that the bayonet may come in contact with his breast, with a full persuasion, that if the horse is killed by rushing on it, he will fall forward and crush him beneath his weight.

weight. This would consequently break down
the rank, and open a passage for the rest of the
cavalry to enter. Thus it is with all the infantry in Europe. The soldier, in kneeling to receive the shock of the squadron, is conscious
also, that he can receive but little support from
the man immediately behind him, and none at
all from the rear rank man, unless he has, perhaps, kept a shot in reserve. The center and
rear rank man also, seeing the feeble barrier
raised by the front rank, to protect them from
the furious charge of the squadron, and sensible
of their own weakness, on account of the short-
ness of their hand weapons, and their inexpert-
ness in using them as such; and feeling altogether the want of mutual support, or protection,
will be more disposed to fly, than attempt a resistance, which they believe to be ineffectual.
The defect of this order, might however be re-
medied, by having the third rank armed with
pikes : " For (as has been observed by M. Saxe,
in his *Reveries*, p. 34) the front rank being
sheltered by the pikes, will take a much surer
aim, and fire with more coolness and resolution
than they otherwise would do ; and the rear
rank being covered by the front will exercise
their pikes with intrepidity, and be capable of

doing

doing infinitely more service, than if they were armed only with firelocks."

It may be doubted also, whether the bayonets of the front rank kneeling, present any thing formidable that might deter the horse from rushing on it: at least it has been said very generally, that the horse has a propensity to rush upon the bayonet, rather than avoid it. It is thus accounted for by an ingenious French military writer: " Of all the arguments advanced in favour of cavalry, the strongest, doubtless, is that propensity of the horse to rush on the bayonet when it is presented to him. I never was witness of this phœnomenon, which I could scarcely believe, but the truth thereof I can no longer question. It is a problem whose solution interests humanity, by the too great disproportion there should be, between the slaughter made by a victorious cavalry, and the loss they would sustain by a defeat: besides this circumstance, though very much in their favour, does not by any means confirm their superiority; it is an advantage they may turn to account, but if they can be deprived of it, its utility ceases. It is true, the bayonet, at the end of a musket, has nothing frightful to the sight of a horse. This weapon presented hori-

nontally

zontally creates no emotion in him, because it appears to him as a reed, differing very little from those he is accustomed to see, in a position no way menacing: it is not surprizing that he advances on it without any apprehension of a consequence he is a perfect stranger to. The disposition and temper of a horse on a day of battle is violent; the whistling of bullets, and the explosion of the powder terrify him: he is stunned by the clamour and cries, and his sides are torn by the spur of the rider; all these are more than sufficient to put him beside himself. His rider prevents his taking any other way to avoid those difficulties, than by crossing this line of men on which he endeavours to precipitate himself. It is probable therefore, that the horse then determines to open himself a passage through this body, which seems no way menacing or terrible: he is ignorant of the effects of the bayonet, and springs on it to escape from the miserable situation he is in, and avoid the dangers that surround him. If this reflection be just, and that the case is really so, the singularity of this phœnomenon disappears, and the means of restraining the action of the horse becomes less difficult."—*Elementary Principles of Tactics, &c. by Sieur B———, Knight of the Military Order of St. Lewis, &c. &c. translated*

by

by an Officer of the Brit. Army, 1761, p.105-6-7. The Sieur B—— mentions as the best method of resisting cavalry, to form the battalions in columns, of not more than twelve men in front, but with a depth of files; and that in the charge, the horse (who is ever ready and eager to avoid danger) will be for gaining the openings on the right and left of those small bodies, and that the least motion of that nature (of endeavouring to make their way into the intervals) will be such a disadvantage to their riders, as must occasion their overthrow, in case they engage; and to determine the horse to that course, the Sieur B—— recommends charging the bayonet with the right hand and foot forward; " as soon," says he, " as the cavalry is got within twelve or fifteen paces of the columns, the soldiers of the front rank, who have already charged their bayonets (p. 18), must raise them with both hands to a level with their hats, the points turned to the enemy, or else they may draw back their hands, and then leaning forward, and supporting their bodies on the right leg, push the bayonet against the nostrils of the horses; and the horsemen the moment the blow is given will be too distant to touch the foot soldier."—*Elementary Principles of Tactics,* p. 110.

NOTE

NOTE (7).

(Remark G.)

" We ought never to have manœuvres performed, without explaining to the soldier the meaning and the benefit that may be drawn from them; by this manner the troops are enticed into discipline, and are ready to perform what is required of them on all occasions."— *Cadet*, p. 53. He quotes from M. Vauban, *Traité de la Guerre en Gen.* vol. ii. p. 117. The soldier should particularly be given to understand the effects of the charge of cavalry, and how he could best oppose them.

" We should always make a battalion conscious of its own force, and of what little consequence an attack of cavalry can be against them, even in an open plain, provided they preserve their order and remain firm. To the shame of the infantry be it, that we cannot destroy the idea among them, that cavalry is the most dangerous to them ; although intelligent people must know, that it is little to be feared, where the infantry is well disciplined, and commanded

manded by good officers." See *Cadet*, as in the last note, p. 54.

The ingenious author of the *Elementary Principles of Tactics*, p. 105, gives the preference to infantry over cavalry; he says:—" Those who consider only the impetuosity of a body of cavalry charging a corps of infantry, ranged in slender order, and without motion, look upon it as naturally impossible for the latter to resist the former; those who reflect on the order of both, and perceive that the infantry opposes more men in front, on the same space of ground, than cavalry can possibly have; who find they have three men in depth to oppose to two horsemen, and almost two files against one ; who sees that each foot soldier has a shot to fire, in a firm position, and on a proper level, against the irregular fire of carabines and pistols, disordered by the motions of the horses; and who, moreover, look upon the horse as liable to fear, and always ready to fly from danger; those, I say, who consider all this, do not hesitate to give the preference to infantry."—*Observations on the Art of War, or Elementary Principles of Tactics, &c. by the Sieur B——*, p. 105.

It may be remarked from what follows, that the author

author above mentioned, as well as many other military writers, have drawn much of their information from Machiavel. What he says on this subject, in the fifth chapter of his *Art of War*, under the head, *The Difference betwixt Men at Arms and Foot, and Upon Which we are most to rely,* observes:—" Again it is frequently seen, a brave and daring man may be upon a bad horse, and a coward upon a good one, and that inequality is the occasion of many disorders. Nor let any one think it strange, that a body of foot can sustain the fury of the horse, because a horse is a sensible creature, and being apprehensive of danger, is not easily brought into it. And if it be considered what forces them on, and what forces them off, it will be found, that that which keeps them off, is greater than that which pricks them on; for that which puts them forward is but a spur, whilst that which keeps them off, is a pike, or a sword. So that it has many times been seen both by ancient and modern experience, that a body of foot are secure, and insuperable by horse. If you object, that coming on galloping to the charge makes the horse rush furiously upon the enemy, and to be less careful of the pike than the spur; I answer, that though a horse be in his career, when he sees the pikes he will stop of himself; and when he feels them

prick,

prick, he will stop short; and when you press him on, will turn either on the one side or the other: and if you have a mind to make the experiment, try, if you can, to run a horse against a wall, and you shall find very few that will do it."—*Fol. Vol.* already quoted, p. 454.

NOTE (8).

(Remark G.)

Our Cavalry Regulations say, "It is in the uniform velocity of the squadron, that its effect consists: the spur, as much as the sword, tends to overset an opposite enemy: when the one has nearly accomplished this end, the other may complete it."—*Cav. Reg.* p. 32.

NOTE (9).

(Remark G.)

How far it may be necessary to practise the horseman so much in exercising the edge of his sword, and depending on it to defeat his adversary,

sary may be doubted; some have gone so far as
to contemn it altogether. " The Romans,"
says Vegetius, " were taught not to cut, but
to thrust with their swords, for they not only
made a jest of those who fought with the edge
of that weapon, but always found them an easy
conquest."—*Vegetius,* b. i. p. 23, English translation.

Machiavel uses almost the same words, and
gives the preference to the thrust instead of the
cut. " Because," says he, " thrusts are more
mortal, more hard to be defended, and he that
makes it is not so easily discovered; and is rea-
dier to double his thrust than his blow."—*Machiavel's Art of War,* fo. edit. p. 455.

Monsieur Guibert probably derives his ideas
from the same sources, aided by his own experience. " The weapons of the brave," says he,
" of the man who wishes to close with his adversary and attack him with success, should be
short, solid, and firm in the gripe. It would
be proper to practise the point in preference to
the edge. This first invented method of fighting is infinitely more destructive; and uniting
both bravery and skill in those nations which
adopt it. To use the point with effect, the man
should

should boldly lay himself open, and mark the spot to level his blow at : but instead of so doing, the custom of using this instrument is to put himself in parade, present a safe guard, now and then risk a blow, and in a very puny manner let his sword fall on the adversary; an effemi nate and inactive kind of defence."—*Gen. Ess. on Tactics,* vol. i. p. 267.

A foreign general officer reasons in the follow-lowing manner on the inutility of using the edge of the sabre, when cavalry are opposed to cavalry. He says, " The horse occupies seven feet and a half in length; the horseman is consequently seven feet and a half from the enemy on horseback : the longest sabre, added to the longest arm, scarce ever extends beyond five feet. I suppose that rising in the stirrups, every horseman, face to face with another, gains a foot and a quarter, still the sabres can scarce cross each other; cuts or scratches of the fingers are the consequence. Is it worth the pains to make dissertations how to give cuts with the sabre, which have, in fact, no existence?"—*British Military Library,* vol. i. p. 276.

And Marshal Saxe totally disapproved using the edge of the sword in preference to the point.
" Their

" Their blades," says he, " must be three square, that they may be effectually prevented from ever attempting to cut with them in action, which method of using the sword never does much execution; they are also much stiffer and more durable than the flat kind: they must be four feet in length; for a long sword is as necessary on horseback, as a short one is on foot." *Reveries*, p. 49.

The contrary opinion has, however, been maintained. " Instead of the straight swords that the dragoons carry commonly in Germany (says a French author) they should have bent ones, like those of the hussars. He admits, that the wound of the straight sword is the most mortal; but that in a battle they cannot be used with so much advantage as the crooked sabre: to prove it, it is only necessary to compare the mechanism of the one with the other. When the horse is in full gallop for the attack, the horseman presents the point of his sword to his enemy; if he pierces him, it consequently stops his horse and his action, to disengage his sword. In this time a dragoon, armed with his bent sabre, would, without stopping his horse or his hand, wound three or four of his enemies, if not mortally, at least sufficiently to disable them

N   from

from the combat; this is what one ought to endeavour to effect in a battle, as much for the
interest as the glory of the Sovereign. In addition to what has been said, the author remarks, that the *cut* of the straight sword is not
so effective as that of the bent one; and that it
is, no doubt, for this reason that the preference
is given to it by all the oriental cavalry; who,
he says, are the best in the world. His words
are, " Au lieu de sabres droits, que les dragons
portent ordinairement en Allemagne; on leur
en donneroit des courbes, pareils à ceux des
hussards. Je sais que les premiers ont le coup
plus meurtrier; mais il s'en faut de beaucoup,
que dans un combat, ils soient si avantageux.
Pour en convaincre; je ne veux que comparer
le méchanisme des uns à celui des autres. Quand
le cheval est en plein galop, pour attaquer; le
chevalier portant la point du sabre à son ennemi,
doit le percer; par consequent arreter son cheval & son action, pour dégager le sabre. Pen-
dant ce tems là, un dragon armé d'un sabre
courbe, aura sans retenir ni son cheval, ni sa
main, blessé trois au quatre de ses ennemis, si
non mortellement, du moins seront ils hors de
combat; c'est ce qu'on doit chercher dans une
bataille; tant pour l'interêt, que pour la gloire
du Souverain. On sait de plus, que le tran-
chant du sabre droit, ne se prêt pas si bien que

le

le courbe; d'où vient sans doute, la preference que fait de celui ci, tout la cavalerie orientale : la meilleure qui soit au monde."—*Le Partisan, ou l'Art de faire la Petite Guerre, par M. De Jeney, p.* 17.

N O T E (10).

(Remark G.)

In a pamphlet, entitled, *Remarks on the In-utility of the Third Rank of Firelocks, &c.* pages 9, 24, and 25. " If it were objected that the third rank of firelocks cannot be dispensed with, as occasions might occur (as in the instance of the front rank kneeling) where its fire might be required; the objection could easily be ob-viated, by the pikemen having each a firelock slung at his back, as our pioneers have." And this is according to the idea of Marshal Saxe : " With regard to my pikes," says he, " if in rough or mountainous places they become use-less, the soldiers have nothing more to do than to lay them aside for the time, and make use of their fusils, which they always carry slung over their shoulders for such purposes."—*Reveries,* p. 39.

They

They would have no occasion for a bayonet, as whenever the hand weapon was required, the pike would necessarily be used in preference, except in very close quarters, in which the short stabbing sword, which every soldier probably should be furnished with, would have the advantage over either of the other weapons, that is to say, the pike, or the firelock with the bayonet fixed on it. There seems to be no good reason why the foot soldier has been deprived of his sword: if the point of his bayonet should be broken, or that in coming down to the charge the bayonet should fly off, as not unfrequently happens, then the soldier with his firelock as a hand weapon is in a very defenceless state, and entirely at the mercy of his enemy. A short, light, bent sword, has been recommended, by M. Jeney, even for light troops, as easily carried, without embarrassing the soldier. " Pour l'armement," says he, " rien ne me paroit moins embarrassant que de donner à un fantassin, outre le fusil & la bayonette un sabre courbe, court & leger ; ce qu'il porte avec beaucoup plus d'aisance," &c.—*Le Partisan,* p. 16.

NOTE

NOTE (11).

(Remark G.)

Several serjeants and soldiers of the Coldstream Regiment, who had been on service, and who attended the trials made in Hyde-Park, were surprized that the men in Experiments No. 3, 4, and 5, could not fire quicker; averring, that they had fired much quicker in action: one of the serjeants was allowed a trial, but the effect was the same; they all agreed then, that if they were allowed to have the cartridges in their pocket, or handed to them as required, that they could easily fire four times in the minute at least; this shows the ill construction of the pouch, and the reason why soldiers are apt to have their cartridges in their pockets instead of their pouches, in their firings. To prevent this practice, the great King of Prussia had it in orders, that " the cartridges shall be always taken out of the pouch, in the firings; no man therefore must stick them either under his waist-belt, or elsewhere."—*Regulations for the Prussian Infantry*, p. 47.

An⋯⋯

Among the other defects of the pouch (which have been attempted to be pointed out, in the pamphlet entitled, *Remarks on the Inutility of the Third Rank of Firelocks, &c.*) may be mentioned, how much it embarrasses the soldier whenever he runs, as in such case he must keep his right hand on it, to prevent its flapping up and down, and losing perhaps all the ammunition. For light troops it has been recommended, " having their cartridges placed horizontal, and covered with wax-cloth, instead of our wooden blocks."—*Military Instructions for Officers detached in the Field, &c.* p. 43.

N O T E (12).

(Remark G.)

It has been endeavoured to be shewn (Note 6, Remark G.), that the front rank kneeling to repel the charge of cavalry is a bad position. The Sieur B——, already quoted, condemns it altogether; as he also does the method of using the firelock as a hand weapon, with the left hand and side forward, instead of the right, as being

being unnatural; for if attacked sword in hand, the soldier must either fight with his left hand forward, or change his position, and being reduced to this necessity (he remarks), clearly evinces the insufficiency of the established method. " The peasant," says he, " when he makes use of his pitchfork, has the right hand before, and the left on the end of the handle. Were he to fight, he certainly would take the same position, without considering that established for the musket and bayonet."—*Elementary Principles of Tactics*, p. 15.

The same author (p. 18) shews in what manner he would have the three ranks taught to charge the bayonet, and use the right hand forward and oppose a barrier to the charge.

" The soldier of the front rank being in the position of recovered arms, must move his right foot two paces forwards, the left hand lowering the firelock gently, until the butt rests against the left thigh, under the watch pocket, the right hand to be shifted immediately, in order to lay hold of it above the swell, and the bayonet to be raised to a level with the right eye.

" The soldier of the second rank must ad-

vance

vance his left foot before the point of the right,
and half level the piece, the butt-end only resting between the nipple of the right breast and
his arm, and the bayonet on a level with the
eye.

The soldier of the third rank is likewise to
advance the left foot before the other, which he
is to move at the same time to the right, in order to half level his piece as the second rank.
This position is gained by a single motion, and
exhibits the soldier of the first rank sheltered
from the enemy's sword by the position of his
firelock, and those of the soldiers of the second rank, the whole length of which is before
him. He is full master of his piece, as it is managed by his right hand, and as he can by that
means exert all his strength against the weapon
of his adversary. He can advance or retreat,
as circumstances require. The front of the
line is in a manner bristled with bayonets, as
three are seen for one; and the bayonets of the
second rank farther beyond the first rank, than
the bayonets even of the first rank in the present posture."—P. 20. Which posture he ex-
plains thus :—" The present mode of charging
the bayonet is by turning to the right, the musket laid horizontally on the left arm, with
the

the lock opposite the pit of the stomach."—
P. 14. This was the mode also practised at that
time by the British army.

The French, at this day, charge bayonets
much as we do, except that the point of the
bayonet is raised as high as the eye; a method
perhaps to be preferred to ours. The only good
reason for making the men of the front rank
kneel, is, that it permits the other ranks to fire
over them: to resist the charge of cavalry, it
might be more effectual, perhaps, for them to
stand up, than to kneel.

N O T E (13).

(Remark G.)

It may seem extraordinary to some, why the
horse should be pricked in the nostril, instead
of having the bayonet plunged into his breast;
but the reason given by the Sieur B—— seems
to be conclusive on this point: " The wounds
of a sharp pointed instrument," says he, " given
suddenly in a direct line, occasion no great

pain

pain for the moment. The horse rushing on the bayonet is in this situation; he is already in the ranks of the infantry before he is sensible of the wound, which is sufficient to throw them into disorder, though the horse and rider should both fall breathless to the ground. The wound meant here (wounding the nostrils) is quite different; its oblique direction tears the skin and muscles; it raises a contusion on the bones, gives instant pain, and reaches far enough to stop the horse, and hinder him from either entering, or falling in the ranks; such a wound is sufficient, I fancy, to stop him quite short." — *Elementary Principles of Tactics,* p. 110.

N O T E (14).

(Remark G.)

Why it is recommended to aim the point of the spear at the face, instead of the body of the man, is fully explained in a small pamphlet alluded to, entitled, *Remarks on the Inefficacy of the Third Rank of Firelocks;* see also *Plutarch's Life of Cæsar and of Pompey.*

NOTE

N O T E (15).

What the French call *impetuosité du choc :* to give this its full effect the idea is, that the cavalry must advance in correct dress, so as to attain the enemy on all his points at the same instant, and overturn him. In the shock of cavalry against cavalry, 'tis not to be supposed that the adverse horses knock their foreheads against each other, much less that their breasts can come in contact, which is what is meant by the *poitrail shock,* or percussion, which is made in the charge against the breast of the adversary's horse; a thing indeed impossible to happen. This is M. Guibert's opinion; he says also, that " in general, when two squadrons are on the charge, the one perhaps does not attain the enemy, and the other is not prepared for it. That which has the less degree of speed, order, and courage, floats in the ranks, is confused, and intirely disordered at the wings; for which reason it either retreats, or else makes but a short and unsoldier-like engagement; but if the two squadrons are composed of well disciplined

men

men and horses, alike exercised, the charge will be thus effected : the ranks mutually close, the instinct of the horses will impel them through the intervals : the men oppose body to body, and the whole is blended in such a manner, that the squadrons pass the one into the rear of the other : therefore, in this *melée*, its success will depend on the skill of the men, and on the activity of the horses."—*Gen. Essay on Tactics,* vol. i. p. 273.

" The strength of cavalry," says an experienced officer, " consists in motion and rapidity; but I cannot think the shock of cavalry must be always decisive. In the campaign of 1762, I was witness of a shock which the greatest part of the Prussian cavalry made upon a superior number of the Austrian cavalry. The consequence was, some hundreds were wounded and taken on both sides ; not a single man lay dead upon the place of action."—*Templehoffe's Seven Years War, translated by the Hon. Col. Lindsay,* vol. i. p. 15.

NOTE

NOTE (16).

(Remark G.)

It may seem extraordinary to many, that in charging the bayonet, the right hand was not placed foremost instead of the left; it is accounted for in the following manner, by the Sieur B——, the intelligent author already quoted :—" The musket," says he, " when first introduced, was nothing more than a missive weapon, which was used with the butt-end against the right shoulder, the right hand behind the touch-hole to apply the match, and the left before to support it. As often therefore as there was occasion for using it, it was natural to fix the hands on the parts described, in order to rest the butt end afterwards against the shoulder and fire ; such was the origin of the method of *poizing the firelock* (the *Recover*), which was rational enough, as long as the musket remained only a missive weapon; but from the time it partook of the qualities of a half pike, by the addition of the bayonet, the ancient position became faulty, and should have been altered: but from the influence of custom and prejudice,

no

no alteration has taken place, and every nation in Europe has been a sufferer. The inventors of the bayonet, accustomed to make use of their fire, continued to look upon it as the principal, and almost the only object, and never once thought of employing this new weapon, but in the most favourable position for presenting.— They concluded, the effect of the fire would sufficiently atone for the difference of dexterity between the right hand and the left. Their error has been faithfully transmitted, and as faithfully followed by all their successors to the present generation. If the origin of the two positions be accurate, the little inclination troops have to come to a close attack no longer remains a mystery. Would so many men who are ready enough to draw their swords, shudder at the sight of a bayonet, if they were generally acquainted with its use? Indeed a weapon whose use is not properly known, is not apt to inspire much confidence. The knowledge of his arms, and their use, should therefore be the first lesson given to a soldier."—*Elements of Military Tactics, &c.* p. 16 and 18.

In Turenne's time the pikemen in all the armies of Europe charged with the left side forward, the butt held in the right hand, so as to

be

be able to push or resist with greater force. And this was the method in which the front ranks of the Grecian phalanx held their pikes in the charge, according to Polybius, as quoted by M. Saxe, *Reveries*, p. 145. And the mode of charging the bayonet, when it was substituted for the pike, was nearly the same, as appears by the following description of the *Charge*, taken from *The Art of War*, published by R. Morden, in the reign of James the Second, p. 12.—" The pike being advanced, *Charge*. Fall back with your right leg, so that the heel of your left foot may be directly against the middle of your right foot. Bring down your pike extremely quick with a jerk, and charge breast high; your left elbow under your pike to support it, always holding the but-end of your pike in the palm of your right hand, and your left foot pointing in a straight line with your pike." This was the position the French called *pique en avant*.

In this position troops could move but slowly, which was probably one reason for inducing Marshal Saxe to alter the position to that now in practice, where a rapid charge can be made with the fixed bayonet.

NOTE

N O T E (17).

(Remark G.)

" Although," says M. Guibert, " all kinds of musket-proof arms are very properly laid aside, yet there is no necessity to disallow of some precautions being used to defend the cavalry against the *arm blanche*, provided they are neither heavy nor embarrassing. It is necessary to 'cover the head with a casque sabre proof, and the shoulders with a three-chained mail fastened to a leathern epaulet. The heads and shoulders of the infantry (he observes) should be covered in the same manner: men having the head and shoulders bare, seek more how to avoid the stroke, than to kill those who are giving it. Let the soldier's head (says he) be protected, and he will fancy that his whole body is safe; we find this kind of illusive instinct exist amongst the most parts of animals."—*General Essay on Tactics*, vol. i. p. 182, 183, 266.

General Lloyde recommends " cuirasses and caps of the strongest leather, for the infantry." —*Rhapsody*, part ii. p. 167.

Probably the General meant *cuir-bouillée* (jacked leather), of which our troopers boots were formerly made, and from which it is presumed they took the name of jack boots.

" The French offered great premiums for any who could make jacked leather musket proof."— *Grose's Hist. of the Brit. Army*, vol. i. p. 103.

" Troops," says Vegetius, " defenceless and exposed to all the weapons of the enemy are more disposed to fly than fight; for it is certain, a man will fight with greater courage and confidence, when he finds himself properly armed for defence." He observes further, "That both the Roman infantry and cavalry was invariably covered with defensive armour; such as helmets, breast plates, greaves, &c. from the foundation of the city to the reign of the Emperor Gratian : that in the decline of the empire and military tactics, their enemies profited by their fatal omission and negligence in protecting the vital parts by the usual military accoutrements." *Lloyd's Rhapsody*, part ii. p. 167.

The English translator of M. Guibert (vol. i. p. 181) remarks, that if the helmet, chain epaulet, &c. was thought too cumbersome for

o   the

the soldier, that the men might be eased " by slinging the casques behind them with a string, and covering their heads with a cloth cap, as was frequently practised by the Romans : the guard *d'épaule* might be rolled up, and tied on the knapsack."

. Not many months since a very fine battalion was seen to go through a field day, with the colours out, and the men with their foraging caps on, to preserve their varnished caps, lest they should be sullied, by dust, or the smoke of the powder. The French cavalry wear helmets ornamented with horse hair; some of the regiments have cuirasses : a regiment so armed (all other things being supposed to be equal) would have a decided advantage over any other cavalry regiment, that was not so well armed. All armour made heavy or cumbersome must be bad, but a plastron or cuirass to defend the fore part of the body, may be made to resist the bayonet, or the sword, and yet so light as to be scarcely felt. Such was that which Plutarch mentions to have been worn by Alexander the Great, which was made of quilted cotton. Marshal Saxe wore a plastron or cuirass of silk stuff, quilted and doubled many times, in the folds of which, cotton was bedded : it was proof, and did not weigh

weigh more than four or five pounds. M. le
Maréchal de Saxe avoit un plastron d'ètoff de
soie piquee en plusieurs doubles que receloient
des lits de coton. Il étoit à l'epreuve, & ne pe-
soit que quatre à cinque livres."—*Esp. des Loix
de la Tactique ou Notes de M. le Marechal de
Saxe, commentées par M. de Bonneville,* tom. 1,
p. 149.

The ancients were very careful of the lives of
their men; honouring him who saved the life
of one of his fellow soldiers, more than him
who killed two of his enemy. It was infamous
for any soldier to lose his shield in battle. The
Greek lawgivers punished the man who threw
away his shield, but not him who lost his sword
and spear."—*Plutarch's Life of Pelonidas.* See
numerous instances on this subject, in the *Target,* p. 36, 37, &c.

Marshal Saxe imputes the laying aside of armour, to nothing but indolence and effeminacy.
"The Romans," says he, " conquered the uni-
verse by the force of their discipline; and in
proportion as that declined, their power de-
creased. When the Emperor Gratian had suffered the legions to quit their cuirasses and helmets, because the soldiers, enervated by idleness,

com-

complained that they were too heavy, their
success forsook them; and those very barba-
rians whom they had formerly defeated in such
numbers, and who had worn their chains during so many ages, became their conquerors."—
Reveries, p. 48.

"I have," says the Marshal, "invented a suit
of armour, consisting of thin iron plates fixed
upon a strong buff-skin, the entire weight of
which does not exceed 30 pounds ; it is proof
against the sword and pike : and although I cannot allege it to be the same against a ball (a
pistol ball), especially one that is fired point-
blank : nevertheless it will resist all such as have
not been well rammed down; as become loose in
the barrel, by the motion of the horses, or are received in an oblique direction," &c. He recommends this sort of armour also on the score of
œconomy. The Roman helmet he prefers, as so
graceful, that nothing can be comparable to it ;
and it lasts, as does the armour, during a man's life.
"Thus," says he, "the dress will be rendered
much less costly, and more ornamental ; your
cavalry will no longer be in a condition to dread
that of the enemy; but rather be fired, from a
sense of their superiority, with an eagerness to
engage them. The Prince who first introduces
this custom amongst his troops, will reap his advantage

vantage from it: for I should not at all be surprized to see ten or a dozen such horsemen attack and defeat a whole squadron, because feat would prevail on one side, and courage on the other."—*Reveries*, pages 45 and 46.

N O T E (18).

(Remark G.)

See *Rules and Regulations*, p. 34.

N O T E (1.)

(Remark II.)

" VOLLIES may be given on a retiring enemy, by the three ranks, the front one kneeling." *Rules and Regulations*, p. 350.

N O T E (2).

(Remark H.)

To these repeated charges of the cavalry on the infantry, cavalry officers might perhaps object,

ject, that it would be forming the infantry at the expence of the cavalry; but even if this were the fact, which might however be doubted, yet, as M. Guibert well observes, " The most important object to be considered is, the forming of the infantry so as to be able to withstand any attack of cavalry, a circumstance which is too universally neglected. It is only since the decay of military discipline, that the cavalry has had, in regard to its charge, the superiority over infantry."—*Gen. Essay on Tactics,* vol. i. p. 188.

M. Guibert, in the above, has probably followed the opinions of Machiavel, who says :— " By many arguments and examples it may be proved, that the Romans in their military exploits had greater estimation for their infantry, than their horse; and that all their principal designs were executed by their foot." See the Second Book of his admirable *Discourses,* chap. 18, p. 355, fo. edit.

In another place, he says, " That nation or kingdom which prefers their horse to their foot, shall always be weak, and in danger of ruin.— Horse should not (he says) be made the strength of the army, but sufficient to second the foot,

for

for they are of great use for scouting, making inroads into the enemy's country, raising contributions, infesting the enemy, and cutting off convoys and supplies of provisions; nevertheless, when they come to *a field fight, which is the main importance* of a war, and the very end for which armies are raised, they are not so serviceable as foot; though, indeed, in a rout, they are better to pursue."—*Machiavel's Art of War,* chap. vi. p. 453.

Many instances are given by the great historian last mentioned, where the infantry attacked the cavalry and defeated them: it is, however, to be observed, that the infantry alluded to by him, as well the ancients as those of his own time, were much better armed, either for attack or defence than those of the present day.

———

NOTE (3).

(Remark H.)

Machiavel, who cannot be too much studied by every military man, giving the preference to foot over horse, among other reasons mentions:

" The

" The foot can get into many places, where the
horse cannot get; the foot keep their ranks better than the horse, and in any disorder are
sooner rallied, and in a posture again, whereas
the horse are, more unmanageable, and when
once out of order, with great difficulty to be
rallied; besides, as it is among men, so it is
among horses, some are high spirited and courageous, others are untoward and dull; and it
frequently happens, that a mettled horse has a
cowardly rider, or a mettled rider a dull horse;
be it which it will, the disparity is inconvenient.
A body of foot, well ordered and drawn up,
will easily be too hard for the same number of
horse; but the same number of horse will have
hard service to break a body of foot, if there be
any thing of proportion betwixt them. And
this opinion is confirmed not only by ancient
and modern examples, but by the relations and
constitutions of legislators, and whoever else
have left any rules and directions for the government of an army? They tell us, indeed,
that at first horse were in greatest reputation,
because the way of ordering the foot was not
known; but as soon as the way of managing
them was found out, and their usefulness was
discovered, they were preferred to the horse."
The Discourses of Nicholas Machiavel, b. ii.
c. xviii. p. 356, fo. edit.

Were

Were the third rank of the infantry now armed with pikes, as has been recommended, they would be invincible to horse or foot, armed as they are at this day.

N O T E (4).

(Remark H.)

This method of breaking down the infantry is not a new idea; for this purpose not only the eyes of the horse were covered, that he might not be intimidated at the sight of the enemy in a charge, but his head was intirely covered with armour, called a chamfraine, which has been already mentioned (Note 8, Remark A).

NOTE (1).

(Remark I.)

FOR this uncertainty of the shot beyond 80 toises, M. Guibert gives many satisfactory reasons :—" The soldier in his hurry and confusion loads hastily, levels worse, trusts to chance in his fire, and is intirely void of that deliberate composure, which would increase his execution." He accounts for this, " by the little pains taken to instruct the soldier in the principles of the fire, adjusting his aim at different distances, &c, and practising very quick firing, which is always but little effective." See for the above, and much information on this truly important subject, *Gen. Essay on Tactics*, vol. i. from page 156 to 176.

Soldiers are not enough instructed in firing at a mark; a target is not the best object to practice them at ; a board the height of a man, and twenty-two inches broad, divided into three equal parts, by red or black lines, would be better: this he should be practised at, making him

take

take his aim at one or other of the lines, according to the distance it was placed at from
him; and it should be explained to him why he
was to aim at one part more than another; but
as this has been dwelt on more fully in another
place, it need not be detailed here. See *Remarks
on the Inutility of the Third Rank of Firelocks,
&c.* 1805.

It may however be remarked, that in practising at the target, there does not appear to be
much emulation among the men; the best shot
may get praise at the moment perhaps, but (in
some regiments) he gets nothing more, not even
an exemption from a duty; there is nothing to
stimulate the soldier to excel in that which is of
the utmost public concern. Perhaps too much
attention is paid to make the soldier fire quick,
rather than fire well.

It is laid down in the Regulations for our Infantry, " That when fire commences against
infantry, it cannot (consistent with order, and
other circumstances) be too heavy or too quick
while it lasts, and till the enemy is beaten or
repulsed."—§. 206, p. 349.

But the judicious author of the British System
of

of Tactics, when he mentions that firing cannot be too quick, while it lasts, supposes the soldier to be perfectly well drilled, otherwise (he well knew) that quick firing could not be very effective.

M. Guibert observes, on this subject, that " those Prussian battalions so famously esteemed for their order and execution, are those whose fire is less galling: their first discharge has precision and effect, because their shot is loaded out of action, and done with more attention and regularity ; but afterwards, in the heat and confusion of an engagement, they load in haste, and are inattentive to the well ramming of their charges."—Vol. i. p. 162.

Another writer of great authority mentions some of the ill effects of very quick firing :— " The precipitation wherewith a solder is at present obliged to load his firelock, does not admit of his *ramming down the charge,* nor of presenting properly, so that his fire may take place.— The barrel of his piece sometimes grows so hot, that it occasions accidents, and perhaps disables him from further service; the barrel may burst, or not throw the ball above thirty paces; in short, the hurry confuses him, he minds not the word of command, nor knows he what he is about.

about. What must follow? Disorder."—*M.*
Bombell's Service de l'Infanterie, p. 37. *Cadet,*
p. 62.

To counteract this precipitancy, the author of
the *Cadet* remarks : " That precious *sang-froid*
so necessary to a military man, should tempt us
to calm that dangerous precipitation, by recom-
mending silence to our soldiers, and to take a
proper time in loading, and wait the word of
command, that they may point their fire where
they wish to execute; this will disorder the
enemy, the other but disorders themselves."—
Cadet, p. 63.

M. Guibert assigns many reasons also why
the fire is not more effective; among others,
that it is always practised at hazard, and its
execution rendered nothing more than merely
mechanical. That in the infantry, there are so
few officers, who understand the construction of
the pieces, and have studied the jet of the mo-
tional bodies which are impelled from them;
that the soldier has no principle given him,
whereby he can adjust himself, for let the distance or the situation of objects be what they
will, he fires at random. One common rule is
however given ; the officers say to the men fire
quick,

quick, and take aim at the middle of your men:
they are likewise told, *level low—the shot will always sufficiently rise*; as if it was possible for
the balls to rise above the line of direction, and
as though there was no law of weight and ten-
dency, to compel all bodies in motion to gravi-
tate towards the earth. Are we after this to be
surprized, if the fire of our infantry is so de-
spicable, so little executive, if in battle, there
being five hundred thousand shot discharged,
that in the field there only remains two thousand slain ?" And he imputes much of this ignorance and deficiency of principle, to the exercise of the target being so much neglected.
See *General Essay on Tactics*, vol. i. pages
161-2-3-4.

In another French work may be found many
valuable remarks on this interesting subject, and
of the means formerly used to encourage the
soldier to become a good marksman, and so obtain a superiority over the enemy. It was for
this reason (says the author alluded to) that
formerly they applied themselves carefully in
our troops, to the means of acquiring that important science: for which purpose, they excited
the emulation of the soldiers, by the hopes of a
recompence that was granted every time to
                                                    him

him who carried away the prize by his skill; this reward was a sword full mounted, a laced hat, or any other thing that the commander of the regiment (maître de camp) gave, or which was purchased at the common expence of the captains, a conduct well worthy of imitation. The words are, " C'est par cette raison qu'anciennement on s'appliquoit avec soin dans nos troupes aux moyens d'acquerir cette science importante, pour cet effet, on excitoit l'emulation des soldats par l'appas d'une reçompence qu'on accordoit, chaque fois à celui qui remportoit le prix par son adresse, cet recompence étoit une épée propre, un chapeau bordé, ou autre chose que le mestre de camp donnoit, ou qui s'achetoit aux frais communs des capitaines," &c.— *Ecole de Mars*, tom. ii. p. 738. The author, M. de Guignard, laments that the practice was in his time disused.

Many judicious observations on this important subject may be found in Tielk's *Account of the War in Germany, from the Year* 1756 *to* 1763 :—" If you compare in a battle the number of musket shots that are fired with their effect, the small proportion that the latter generally bears to the former is quite astonishing. The original fault may be traced to the method

of

of training the troops, in which more attention is often paid to the parade than to what is generally essential from its immediate relation to actual service. The making every soldier of a battalion practise at a target twenty times a year, would be of more real use, than if the battalion was to fire a thousand parade rounds in the closest and most precise manner possible: The mere popping altogether is certainly a trivial consideration, when compared to that essential point of aiming with exactness."—*Note by the Translator, Col. Craufurd.* See vol. i. pages 153-4-5.

N O T E (2).

(Remark I.)

" To hurry and bring up troops to the attack in imperfect order, is to lose every advantage which discipline proposes, and to present them to the enemy in that very state to which after his best efforts he has hoped to reduce them." *Rules and Regulations,* p. 333, edit. 1803.

NOTE

N O T E (3).

" The musket in itself," as the Sieur B——
óbserves, " is but a missive weapon; with the
bayonet fixed, it is a missive weapon, partaking
in some measure of the nature of a half pike.
The manner of using it as a missive weapon
should be different from that where it is employed as a half pike. It is therefore necessary
to give the soldier a mixed position, so that he
may, at a single motion, shift to either, as circumstances shall require."—*Elementary Principles of Tactics*, p. 11. See also Note (12), Remark G.

" Our infantry," says M. Guibert, " have
two united properties, since the musket armed
with the bayonet is at the same time a missive
and a hand weapon: as a musket it is serviceable in every act of mission, and a weapon of
repulse wherever the bayonet is employed in actions of the shock. Here I cannot help re-
marking, how much the musket, armed with
the bayonet, strikes me as a weapon superior to
any of the ancients. Respecting this weapon,

P   im-

improvements might be made, and greater utility drawn from it than is imagined. A kind of fencing might be learnt, so as to be able to use it with greater effect, &c.—*Gen. Essay on Tactics*, vol. i. p. 117.

The ingenious author of the *Military Dictionary* (second edition), under the head *Bayonet*, mentions, " That another French author (Mauvillion) asks, how can any man tilt or fence with so cumbrous an instrument, and so difficult to be handled, as the firelock? The utmost that could be done would be to make one thrust, and yet that could not be effected with any degree of ease or certainty."

M. Mauvillion seems to have declared his opinion rather too hastily, in asserting, that *the utmost* which could be done would be *to make one thrust, &c.* Experience has proved that notwithstanding its weight, the modes of attack and defence with the bayonet may be reduced to a system similar to those of the broad sword and quarter staff; and that great advantage might be gained, in single contest, by a person previously instructed in such exercise. How far it might be politic to adopt a mode of fencing by a body of men in line, will perhaps admit of
                                                        much

much argument. Those who contend strongly for its advantages to an individnal, are yet inclined to admit, that the success of a charge with the bayonet depends principally on the rapid and firm advance of a body of men, which striking terror into those waiting to receive the shock, produces wavering and disorder in their ranks; and that this effect would be much weakened, if not entirely lost, by a cautious and slow advance in an attitude rather calculated for defence than attack.

It may be remarked here, that the present method of charging the bayonet was proposed to the King of Prussia by Marshal Saxe, in 1753. His Majesty immediately adopted it; before that, they used to march to the charge with the firelock rested on the left arm. See *Esp. des Loix de la Tactique, &c. &c. par M. de Bonneville, tome premiere,* p. 150.

The British charged bayonets in the same manner; and adopted the present alteration after them.

It may be observed here, that bayonets were not known in the wars between the Imperialists and Turks, in 1663-4. Not long after this pe-

riod,

riod they were used, but they screwed into the muzzle of the firelock: they were at first called daggers. The author of the *Target* says, in his Introduction, p. 25, that the bayonet was invented by the people of Malacca, long before it was known in Europe.

———

N O T E (4).

(Remark I.)

" The *attack fire*, or fire in advance, was (says M. Guibert) introduced by the great Frederick into his armies; as he said, that infantry should never be led to the enemy without firing."— *Gen. Essay on Tactics*, vol. i. p. 171.

In the King of Prussia's General Orders issued after the battle of Molwitz (1741), it is mentioned, " When the army, in close order, shall come within 600 paces of the enemy, then, in order to familiarize the soldiers to the fire, and to blind them with regard to danger, they shall begin to fire regularly by platoons."—See *Simes*, p. 94.

This

This regulation did not say much in favour of the valour of his soldiers; national prejudice apart, such reasons cannot apply to British soldiers, who, calm and unmoved, can look danger in the face under its most terrifying aspect. And notwithstanding the high authority of the great King of Prussia, M. Guibert has given reasons that must be conclusive, against making use of the attack fire : " Either insurmountable difficulties," says he, " separate me from the enemy, or else there is a possibility to join them. In the first case, the action is necessarily rendered an engagement of the infantry in steady order. Secondly, I conceive it proper to advance without stopping and firing. To march in firing, or to halt to put it in execution, would be retarding the movement, it would be receiving more fire from the enemy than what it receives from you ; besides a more destructive one, since that of the enemy's, who I conceive either to be firm (halted), or else well posted, would be far more lively and better pointed."

This fire was made sometimes by the men of the center and rear ranks keeping up a continued discharge in marching. " This," as M. Guibert observes, " throws themselves and step out of all manner of form." The Prussians also

call

call *attack fire*, that which consists " in com-
bined and alternative discharges of platoons,
divisions, and half or whole battalions; the part
of the line that has fired, marches in quick time,
and the other which has not in slow. What is
likewise contrary to all military views, is, the
men, by practising this kind of fire are rendered incapable of advancing. As their fire
may possibly attain the enemy, so the enemy
may in return attain them likewise. What military man, in reflecting on these principles, will
not find that this line would soon be shattered,
rendered a complicated heap of consternated
soldiers, broke into intervals, productive of a
strange variety of unequal steps in the march,
and confused by the perpetual shouts of commanders? Would he not likewise say, that this
is all impracticable in war, that the enemy's fire
would endanger the commanding officers, by
which means the alternative order of the fire
would subside; the troops in a mechanical manner, natural on those occasions, would be thrown
in the intervals, and should it so happen, that
these intervals were confounded together, and
their situation changed, the line would no longer
retain its uniformity, but be turned into a confused and disordered mass."

M. Guibert

M. Guibert means to lay it down as a principle, " That there are no fires but the fires in steady order, and *pied firme*, which are practicable, or can be reduced to a certainty in war." In another place he observes, " If you are retiring, every action of the fire would be again ineffectual, and ill placed, for your movement would be retarded, you would be losing time, and deviating from your object, which would be the endeavouring to get out of the enemy's reach, and to gain a post, where you may rally upon, and recommence the battle. I therefore make it my general maxim, *not to fire but when you can no longer continue your movement;* for whether you attack, whether you retire, or pursue a flying enemy, advancing is your chief object, and the only thing which insures you success. In short, by making use of the fire in this manner (the attack fire), it would be losing that decisive advantage of assurance, which expeditious and daring movements inspire troops with: besides the enemy in seeing you advance upon them, in spite of their fire, would be alarmed and give ground." And his English translator adds, in a note, " That by this intrepid cool movement the English troops, last war (1759), gained many signalized

advantages

advantages over the French."—*General Essay on Tactics,* vol. i. p. 171-2-3.

The author of the Rules and Regulations does not appear to be much in favour of the *attack fire,* or fire in advance; at least he deems it incompatible with any thing like a rapid movement. His words are, " When firing in line advancing, the march must be very slow, the line must be preserved, and the officers must take care to point out the supposed object of attack, and see that the men direct their fire to it."— *Rules and Regulations,* p. 280.

Notwithstanding that the great Frederick introduced the *attack fire* into his armies, yet on some occasions he would dispense with it.— Speaking of the oblique order of battle, he observes:—" We must rank battles of this kind among the best, always remembering to attack the weakest. On these occasions I would not admit the infantry to fire, for it only retards the march. The most certain way of insuring victory, is to march briskly, and in good order, against the enemy, always endeavouring to gain ground."—See *King of Prussia's Instructions to General Officers,* p. 128.

On

On the subject of beginning to fire at a great distance, Tielk has the following remarks in his *Account of the War from* 1756 *to* 1763 :—" It is better to allow him (the enemy) to approach within a good shot. The first round is most essential, and must not be given at random; if executed as it ought to be, with great effect, it tends much towards abating the enemy's courage, and diminishing his eagerness to advance.

Is it not absurd and ridiculous to begin firing with musketry at the distance of five or six hundred paces ? And yet I have often seen it done. Many assert in defence of this, that you should engage the soldiers' attention, in order that they may not have time to reflect on the approaching danger, and become faint-hearted. It would, in my opinion, have, in most cases, a contrary effect, &c. It would intimidate the defendants, &c. in the same proportion as the assailant would be encouraged by it. By superfluous and unnecessary firing you may sometimes be reduced to the melancholy situation of not having ammunition when you are most in want of it."

———

NOTE.

NOTE (5).

(Remark I.)

That able tactician the great King of Prussia was of opinion, that troops, though well posted, should rather quit that position, and meet the enemy, than wait to be attacked:—" If troops," says he, " be even well posted, they must quit this post instantly to march against the enemy; who instead of being allowed to begin the attack, is attacked himself, and sees all his projects miscarry. Every movement which we make in presence of the enemy without his expecting it, will certainly produce a good effect." See *King of Prussia's Advice to his Generals*, p. 107.

NOTE (6).

(Remark I.)

Marshal Saxe says, " Had the last war continued some time longer, the close fight would certainly have become the common method of engaging; for the insignificancy of small arms

began

began to be discovered, which make more noise than they do execution ; and which must always occasion the defeat of those who depend upon them."—*Reveries,* p. 19.

And in another place he observes, " That battles would be yet decided with the naked steel, that the fire of small arms would become every day of less consideration, and that cannon and the bayonet would decide the battle."

M. Guibert, observes on this subject (vol. i. p. 146), " That were the fire of troops better pointed, as it would then become more destructive, so it would oblige them to come sooner to close quarters to dispute victory with the naked steel, which is the only mode of fighting applicable to true courage, activity, and skill."

The close fight seems peculiarly well adapted to the determined courage, and superior bodily strength of the British, compared (generally speaking) with that of the French soldier.

NOTE

N O T E (7).

(Remark I.)

Monsieur Guibert, in his excellent work, *L'Essai General de Tactique*, vol. i. p. 130 and 142, has some observations on the little use that is made of the bayonet:—" Whether," says he, " it be owing to custom, or the dwindling away of courage, I know not; but at this day, bodies of infantry seldom make use of the naked steel; and when they march to the charge, they seldom meet near enough to cross or join bayonets together." After mentioning that the soldier is too much familiarized with this weapon, and almost always unnecessarily armed with it, he says, " The display of this weapon, on decisive occasions, would have a terrible and imposing appearance; it would be similar to the crimson standard of the ancients, the signal of death and carnage. From the German infantry we have acquired this custom of always carrying the bayonet fixed; and one circumstance is singular, for ever since they began first to wear it, they have intirely neglected its use."

But it must be recollected, that at first, when the fusil was used only as a hand weapon, that the bayonet, or dagger as it was called, was

screwed

screwed into the muzzle of the piece; and when used as a missive weapon, the bayonet was unscrewed and returned into the sheath; but when the method of fixing the bayonet in the present manner was found out, the firelock in itself united the two properties of a hand and missive weapon; which gives it a decided superiority over any weapon of ancient or modern invention, that we are acquainted with. And unless the soldier was perpetually used to it at the end of his musket, in his general training and exercise, he would never be able (when circumstances required his using the musket as a missive weapon, and at the same time for his own safety keeping the bayonet fixed) to load with the quickness, nor level with the exactness required. And the reason why it was not more used, was not so much owing to the diminution of courage in the soldier, as to the different system of tactic adopted by the commander. That was the fashion of the day, which seems now to be inclining the other way, and the prejudice in favour of the hand weapon appears to be gaining ground; but as it must be desirable in most cases to have the soldier move to the charge with his piece loaded, so the only difficulty appears to be the preventing his using the missive weapon till ordered.

" It

" It is inconsistent," says Marshal Saxe, " for one body of troops to make use of two different ways of engaging at once: they must of necessity therefore either proceed at once to close fight, or depend altogether upon their firings; and whenever the former method is to be put into execution, the latter must be laid aside, to which, on actual service, men can hardly be reconciled; nothing, in general, being more difficult than to prevent their firing when they approach near their enemy; of which he gives the following as a striking instance:—

Charles the Twelfth, King of Sweden, intending to introduce amongst his troops the method of engaging sword in hand, had frequently mentioned his design to his officers, and it was likewise made known to his whole army: accordingly at the battle of ——— against the Muscovites, he hasted to the head of his regiment of infantry, the moment it begun, and made them a fine harangue; immediately after which he dismounted, and posting himself in the front of the colours, led them on to the charge; but as soon as they came within about thirty paces of the enemy, the whole gave fire, notwithstanding his presence, as well as his positive orders to the contrary; and although he routed the enemy,

enemy, and obtained a complete victory, yet he was so piqued, that he passed through the ranks, remounted his horse, and rid off without speaking a single word."—*Reveries*, p. 32.

———————

N O T E (1).

(Remark K.)

Uneven and wet as the ground was at Saumarez Miles (or Common), yet it was the most level that could be selected for the purpose, between Fort Henry and St. Héliers, a distance of about four miles : and it happened very well after all, that the experiment was tried in such ground, as it formed a contrast to that in Hyde-Park, where it was perfectly level, and free from every sort of impediment.

———————

NOTE

N O T E (1).

I allow," says Captain Tielk, in his *Field Engineer*, " that a field piece may be fired six times per minute; I know that it is possible to fire oftener: but will not the excessive hurry prevent your taking aim, and consequently doing any damage? Not more than ten in one hundred shots take place in an action, even where the men have time to point the gun, with coolness and perhaps with safety.

M. Guibert observes, " That it is necessary to fire slow, to be able to point with certainty, and progressively increase the smartness of the fire, according to the diminution of the ranges (*General Essay on Tactics*, vol. i. p. 338);" because when the object is near, the gun being at a point-blank or horizontal level, the necessity of taking aim must cease.

It may be observed, that Captain Tielk says, " If a cannon is fired at the elevation of one degree, it may be considered here as horizontal,

without

without impropriety."—*Tielk's Field Engineer*, vol. ii. p. 226, *Hewgill's Translation*. See also for point-blank and random ranges, *Le Blond's Elements of War*, p. 23.

NOTE (2).

(Remark L.)

Among other experiments the different ranges of grape shot were meant to have been tried. Every one knows that nothing more appals the soldier than the idea of being exposed to the fire of grape shot; but what the extremest distance is, at which, fired from a six-pounder, it could reach, and how wide it would scatter at that distance; or at what distance, from its departure from the gun, it begins to scatter; and to what extent it spreads at each of these distances, were questions, to which (when put to those who might be supposed to be well acquainted with the subject) general or unsatisfactory answers had been given. No conclusions could therefore be safely drawn, without obtaining some data to proceed on; and therefore a plan was

sub-

submitted to Lieutenant-General Gordon, and obtained his approbation; but circumstances intervened that prevented the experiment from being made.

NOTE (3).

(Remark L.)

The quotation is from Colonel Templehoffe's *History of the Seven Years War, translated by the Hon. Col. C. Lindsay,* vol. i. p. 3.

NOTE (4).

(Remark L.)

This experience might be gained in a **great measure** by opposing first small and then large bodies of troops to each other, and commencing a sham fight, in any sort of country, where the commanders on each side (under the direction of

of a skilful general, well used to the manœuvring of troops) might have opportunity of
displaying their talents, by their method of attack, choice of positions, ready applications of
the *coup d'œil*, and skilful retreat; so that when
they came to real action they might not be at
a loss or confused by the manœuvres of the
opposed party, or by their making the counter
movement.

Monsieur Jeney observes, that " it has always been lamented, that men have been
brought on service, without being informed of
the uses of the different manœuvres they have
been practising; and having no ideas of any
thing but the uniformity of the parade, instantly
fall into disorder or confusion, when they lose
the step, or see a deviation from the straight
lines they have been accustomed to at exercise.
It is a pity to see so much attention confined to
shew, and so little given to instruct the troops in
what may be of use to them on service."—*Military Instructions for Officers detached in the
Field*, p. 53.

The young officer also, if attentive and possessing any taste for his profession, will, by these

sort

sort of exercises, find his ideas expand, till he comprehends almost the business of a campaign and the whole circle of his duty. If money or interest, or both, has pushed him prematurely into a higher rank than his length of service could justify, he might (if a man of sense) reflect, that by what he sees before him in the sham fights skilfully conducted, that his eagerness to get forward and throw the veteran far behind, has placed him in a situation the most delicate, and that having command, without military experience, he might by forming a wrong judgment sacrifice the lives of brave men to no purpose; risk the loss of a battle in which he was engaged, and on which the fate of his country might perhaps depend. It has been therefore well observed by a French officer of great professional knowledge, " That no person should ever take the command of troops in time of war, if he has not sufficient capacity to act properly on an occasion without advice; or if he follows too hastily, or is too much attached to his own notions. It were to be wished (says he) that all those who are so solicitous and eager for preferment in the army, conceived all the importance of the affairs which they are so ready to take the direction and charge of. They would

certainly

certainly be more modest in their requests, did they but feel part of the sorrow and remorse which torture the humane and good-natured man through life, after having, by ignorance or neglect, occasioned any of those melancholy disasters that constantly happen in war. But alas! many are too ignorant to be sensible of it: they only accuse fortune, and think of taking their chance again."—*Elementary Principles of Tactics, by the Sieur B——, p.* 213.

N O T E (5).

(Remark L.)

Practice without theory, as a French author well observes, gives but faint ideas, often false, and subject to many errors. Theory without practice is more dangerous still, because it reasons only on vague principles, and ideal plans; whereas practice forms no calculations but on real grounds, and is aided by experience. But when theory and practice are united, they bring each other to perfection: thus it may be laid down, as a certain principle, for every military man desirous of instructing himself in the art

of

of war, that he ought to join theory to practice; because that without this double instruction he can never calculate justly, he can only combine small or trifling objects, still he will often deceive himself, and he can never execute projects on a grand scale.

———

N O T E (6).

(Remark L.)

The distance at which some young and inexperienced officers think it necessary to keep the soldier, and the contemptuous manner in which they commonly address them, must unfortunately preclude this sort of intercourse.— Some think it beneath them even to speak to a soldier with common civility, or to return his salute though he were a sentry, as may be seen by the ungracious manner in which this indispensable part of an officer's duty is performed; the hand is whipped up to the corner of the hat as if the arm was seized with a spasm, and as suddenly dropped down as if the tassel had been a coal of fire that burnt the young gentleman's fingers. The merits of such officers, indeed,

the

the old soldier easily appreciates; but the treat-
ment he meets with from officers who know their
duty, and have the good of the service at heart,
is very different; such know the value of a good
soldier, and treat him as a friend. To young
officers, as here described, who think it deroga-
tory to their dignity to speak kindly to a soldier,
it might not be improper to mention (if we may
be allowed to take a lesson from our enemy),
that a French officer in the old or new school
would not deem it beneath his dignity, however
high in rank, to say *mon camarade*, instead of
you Sir, or *you fellow*, when speaking to a soldier, nor of taking off his hat in a solemn and
graceful manner, when returning the salute of a
soldier, particularly of a sentinel.

Of the great Turenne it was said, that he ad-
dressed the commonest soldier with a " noble
familiarity: by condescending to them, without
debasing himself, being familiar with them without losing any thing of his dignity; he tied to
him, by the bands of affection, men who are
commonly restrained by nothing but the fear of
punishment: a reproof from him was the greatest
chastisement, and his approbation the reward
they most coveted."—*C. Ramsey's History of
the Viscount de Turenne*, vol. ii. p. 491.

The

The following anecdote may shew what opinion that great man entertained of a good soldier:—" Intending to deceive the enemy, he encamped at Marlen, 7th October, 1674, where however he did not intend to remain. He set the men to work at intrenchments at the head of his camp, and all the army believed that he intended to wait the enemy there. As he was visiting the works, he observed an old foot soldier resting himself; the Viscount came up to him, took him aside, and asked him why he did not work: the soldier answered him, smiling, *It is, General, because you will not stay long here.* Turenne by that saw his penetration, gave him money, desired him to keep the secret, and soon after made him a lieutenant."— *History of Turenne,* as above, p. 447.

In another instance, somewhat similar, the Viscount after conversing some time with a corporal on a position that he meant to take, the design of which the corporal penetrated; he was so pleased that he made him a captain on the spot, who filled the post for many years with great honour."—*L'Ecole de Mars,* l. iv. p. 695.

NOTE

N O T E (7).

This steady courage, as characteristic of the English, one of the most enlightened of the French military authors is ready to allow : his opinion of the French and English valour may not be thought out of the way to mention here. Speaking of the military of the different powers of Europe, and how they were constituted, he says of the French, " They had at that time, as at this day (the middle of the 18th century), their first moment of force and impetuosity : that shock, which at one instant nothing can oppose; and in another, the most slight obstacle can repulse; an inconceivable intrepidity of courage, which at times is able to surmount every thing, and a panic very often carried to the greatest excess of weakness." Speaking of the English, he says, " The English had no tactic, very seldom good generals; but an order which accorded with their arms; a courage little capable of the offensive, but difficult to shake. An historian, in reciting the battles of Verneuil, Cressy, and Agincourt, says, that they waited while the ignorance and impetuosity

of

of the French came and dashed themselves to pieces against their firm bravery and cool blood."—*General Essay on Tactics*, vol. i. pages 101-2.

The whole of the chapter from which the above extracts are made, that treats of " The Influence which the Genius, Government, and Arms of a Nation, has over the Tactic," is well worth the attention of every military man.

General Lloyd, who knew the English well, speaks thus of them, in his preface to the *History of the Seven Years War:*—" The English are neither so lively as the French, nor so phlegmatic as the Germans; they resemble more however the former, and are therefore somewhat lively and impatient. If the nature of the English Constitution would admit some degree more of discipline, a more equal distribution of favors, and a total abolishment of the buying and selling of commissions, I think they would surpass, or at least equal any troops in the world."

Machiavel seems to impute the falling off of the French after their first charge more to want of discipline, than any failure of courage;

rage; these are his words: — " This puts me in mind of what Livy says, in many places, of the French; that in their first attack they were more fierce and daring than men, but afterwards more fearful and pusilanimous than women. And many people enquiring into the cause, do attribute it to the peculiarity of their temperature and nature. I am of opinion, that there is much of that in it; yet I cannot but think but that nature which makes them so furious at first, may be so invigorated and improved by art, as to continue their courage to the last."

The curious reader may find the whole of that great man's reasoning on this subject, delivered near three centuries ago, in the Third Book of his *Discourses on the Decade of Livy,* ch. xxvi. under the head of " *The Reasons why at the first Charge the French have been, and still are accounted more than Men, but afterwards less than Women.*"

NOTE

N O T E (8).

(Remark L.)

See *Informations and Instructions for Commanding Generals and others,* p. 12; where speaking of the enemy, it is said, he must be attacked " not only with fire, but with the National Weapon, the Bayonet; or with a pike or pitchfork, weapons equally dangerous."

N O T E (9).

(Remark L.)

Remarks on the Inutility of the Third Rank of Firelocks, &c. and the Propriety of forming the Third Rank of Pikemen, &c.

N O T E (10).

(Remark L.)

See *Grose's History of the British Army.*

" In 1703, the battalions in France had not more than a fifth part which carried pikes, the rest muskets and fusils. There is no mention of any pikes at the famous battle of Blenheim, which was in 1704; though the two armies consisted of Imperialists, English, Danes, Dutch, Prus-

Prussians, Hanoverians, Hessians, French, and
Bavarians."—*Introd. to the Brit. Target*, p. 25.

N O T E (11).

(Remark L.)

M. Guibert condemns severely the practice
in the French armies of training the soldier
only to handle his musket, and to keep, as he
says, " for three hours the most painful atti-
tudes, contrary to the natural mechanism of the
body," and not instructing him in other sort of
exercises, better calculated to make the body
vigorous and robust. " For these reasons (says
he) a march in the smallest degree forced,
frightens and overcomes him; he is stopped by
a rivulet; four days of pioneering work dis-
heartens him, &c." He laments their filling up
his time (that could be so much better employed) with tormenting and odious rules of
discipline. " They have established a regulation
(says he) which obliges them to pass three hours
every day in dressing and tricking themselves
out to the greatest advantage; converting the
soldier into a frisseur, polisher, and varnisher;
in a word, every thing, excepting a warrior.
And what is the result of this superfluous, yet
painful

painful life—of those exercises which are, for
the most part, practised in a sedentary manner,
and under the supine shade of idleness? The
result is this; a soldier who has served ten years,
having lost all his activity and aptitude for the
use and exercise of his body, is necessitated to
make himself either an artist, footman, or beg-
gar." And he goes on to observe, that by ex-
changing those frivolous occupations for more
hardy and laborious exercises, the soldier, on
quitting his vocation, would then without dif-
culty retake his spade, or return to the plough.
General Essay on Tactics, vol. i. p. 113-14.

But in all this, M. Guibert says, that he only
attacks the severe rules of the *tenue* or army regulation, carried to an excess, and not the *tenue*
itself. Perhaps the time of our soldiers is too
much occupied in these sort of sedentary employments. In no part of our regulations for
the army is this excess of polishing, varnishing,
and burnishing enjoined. All that is, or perhaps ought to be required of the soldier, is to
have himself and appointments perfectly neat
and clean; and above all, his musket to be in
proper firing order: it is well known, that burnishing the firelock soon renders it unservicea-
ble.

NOTE

N O T E (12).

(Remark L.)

The words are, " Ils sont plus endurcis a la
fatigue qu'ils sont habitués à manier à fer, à
creuser des fosses & à porter des fardeaux."—
Veg. 1. i. ch. 3. *Turpin's Mcm. de Montecuculi,*
p. 63.

N O T E (13).

(Remark L.)

If authority were wanting, in confirmation of
what has been advanced in the Remark, Machiavel may be safely quoted. His words are,
" Though they be never so well chosen, and
never so well armed, soldiers are carefully to be
exercised; for without exercise they are good for
nothing. And this exercise ought to be three-
fold; one is to inure them to labour and hard-
ship, and make them dexterous and nimble;
another to teach them how to handle their arms;
and the third to teach them to keep their ranks
and orders in their marches, battles and en-
campments: which are three great things in an
army.

army. For if an army marches, is drawn up well, and encamps regularly and skilfully, the general shall gain reputation, let the success be as it will. Wherefore all ancient commonwealths provided particularly for these exercises by their customs and laws, so that nothing of that nature was omitted. They exercised their youth to learn them to be nimble in running, active to leap, strong to throw the bar, and to wrestle, which are all necessary qualities in a soldier; for running and nimbleness fits them for possessing a place before the enemy; to fall upon them on a sudden in their quarters, and pursue with more execution in a rout: activity makes them, with more ease, avoid their blows, leap a ditch, or climb a bank: and strength makes them carry their arms better, strike better, and endure the shock better; and above all to inure them to labour, they accustomed their soldiers to carry great weights, which custom is very necessary; for in great expeditions it happens many times that the soldiers are forced to carry (besides their arms) several days provisions, which without being accustomed to labour, would be more tedious to them; and by this, great dangers are many times avoided, and great victories many times obtained."—*Machiavel's Art of War*, c. vi. fol. edit. p. 455.

See

See also the preliminary chapter of the *General Essay on Tactics,* where M. Guibert observes, " that the present instruction of troops is inconsistent and ridiculous : it consists only in the manual exercise, and a few manœuvres, that in regard to the operations of war, are for the most part complicated and useless. The sóldier (he observes) should be familiarized with the representations of things similar to those that are practised in real service : he should transport materials of different kinds relating to war; be employed in digging the earth, exercised in forced marches, swim across rivers, and work with alacrity and skill in every part of an intrenchment; be taught how to cut and fix a palisade, the placing a ladder, &c. &c."—Vol. i. p. 110.

N O T E (14).

(Remark L.)

See on this interesting subject, *General Essay on Tactics,* vol. i. p. 250, where M. Guibert shews the method, and points out the propriety of exercising troops relatively to the ground.

NOTE (15).

(Remark L.)

M. Guibert remarks, " that in the interval of his duty, by way of recreation, the soldier should be permitted to indulge himself with those amusements the best adapted to promote his vivacity and strength of body."—*General Essay on Tactics,* vol. i. p. 109.

———

NOTE (16).

(Remark L.)

No proper opportunity should ever be neglected of having lads taught to swim. Speaking of the manner in which the youth of a country should be exercised, Machiavel says, " And to these exercises (mentioned in Note 12, Remark L.) I would accustom all the youth in my country, but with more industry and solicitude, those exercises which are useful in war, and all their musters should be on idle days. I would have them learn likewise to swim, which is a very useful thing; for they are are not sure

of

of bridges wherever they come, and boats are not always to be had; so that your army not knowing how to swim, is deprived of several conveniences, and loses many fair opportunities of action. The reason why the Romans exercised their youth in the *Campus Martius*, was because of its nearness to the Tiber, where after they had tired themselves at land, they might refresh, and learn to swim in the water."—*Machiavel's Art of War*, b. ii. ch. vi.

The following instances may serve to shew the utility of having soldiers taught to swim.— Plautus, the Roman general, in the reign of Claudius, gained a great battle over the Britons, under the command of Caractacus their King, principally by means of his swimmers. "The Britons pursuant to their first design (of not giving battle), retire beyond a river (supposed to be the Severn), where they encamp in a careless manner, imagining the Romans could not pass without a bridge; but Plautius had in his army some German soldiers, that were used to swim the strongest currents. These soldiers, though few in number, swimming the river in their arms, so astonish the Britons, that they quit their post," &c.—*Rapin's Hist. of England*, vol. i. p. 13, fo. edit.

In

In the year 78, towards the latter end of the reign of Vespasian, " Agricola, the Roman general, in his first campaign against the Britons, attacked the Isle of Mona (Anglesea), which the Romans had been forced to abandon, though he wanted flat-bottomed vessels for the expedition; he ordered a choice body of auxiliaries, who were acquainted with the shallows, to swim over; which they performed so dexterously (being trained up, by the custom of their native country, to manage, in swimming, themselves, their horses, and arms), that the inhabitants, astonished at the sight, and never suspecting any such thing, surrendered the island up to the Romans, without obliging them to draw a sword."—*Rapin*, as above, vol. i. p. 17.

Montecuculi, speaking of the bravery of the Turkish troops, when he commanded the army of the Imperialists against them in the Hungarian wars, in the years 1663 and 1664, says, " I have seen them throw themselves two different times into the Muer, with their sabres between their teeth, and once in the Raab, to swim across them in our presence, which rendered less surprising the action of those brave Spaniards, who in the time of the Emperor Charles the Fifth, attempted the passage of the Elbe, by swimming,

with

with their swords in their mouths. The words are, " Je les ai yu se jetter dans le Muer par deux fois, le sabre entre les dents, & une fois dans le Raab pour le passer à la nâge en notre presence; ce qui rend moins surprenante l'action de ces braves Espagnols, qui du tems de Charles V. tenterent de meme de passer l'Elbe à la nâge, leurs épées dans la bouche."—*Mem. de Montecuculi*, l. ii. c. ii. p. 261.

Many more instances might be produced to shew the utility, indeed the necessity of soldiers being able to swim. The French have availed themselves well, in several instances, of having good swimmers in their armies.

Leave is taken here to mention, that the ingenious author of the *Military Dictionary* (Major James), to whom the officers of the British Army, the junior ones in particular, are so much obliged, as well for that excellent work, as for his other valuable military writings, where their best interests, improvement, and information, appear to have been the objects of his unwearied attention, has, in the last edition of his Dictionary (under the several heads of *Nager*, *Natation*, and *Swimming*), treated this important subject in a manner at once interesting and satisfactory.

NOTE

N O T E (17).

Every one must allow, that exercise is good for health : standing on a parade for an hour, or more perhaps, most of the time motionless, cannot be conducive to health. The soldier carries his musket always on the same arm, and its weight resting on the same hand, and though moving about at a regulated pace (which may be called exercise for the legs), but no one can say that it is an exercise calculated to brace his nerves, or make him vigorous, or robust ; yet this is all the exercise he gets, and when over, he returns to his sedentary employments of varnishing, polishing, and burnishing, until he again stands on the parade, with every part of his appointments correctly put on, and clean to a degree that can stand the test of the most critical examination. Every military man must delight in seeing a soldier clean, and his appointments at all times in high order. The only thing to be considered is, whether the mode of his exercise, as well as the length of time spent in* polishing and burnishing, does not tend to make the soldier effeminate, and prove injurious to his health. More instances than one has been seen, where
soldiers

soldiers, in good quarters, with no particular
hardships to complain of, yet near an eighth, or
a tenth of a battalion, composed mostly of
young men, have been at once on the doctor's
list, and not many with venereal complaints.—
The sort of exercise that probably a soldier ought
to be sometimes practised in, and at other times
permitted to indulge in, has been shewn ; and
this is the sort of exercise that Machiavel al-
ludes to, when he mentions it as the best pre-
ventative of disease in an army. His words are,
" A general will find himself over-laid when he
has to contend with an enemy and a disease:
but of all remedies, nothing is so powerful as
exercise, and therefore it was a custom among
the ancients to exercise them continually. Think
then of what importance exercise is: when in
camp it keeps you sound, and in the field it makes
you victorious."—*Machiavel's Art of War*, b. vi.
ch. vii.

Printed by C. Roworth, Bell Yard, Temple Bar.